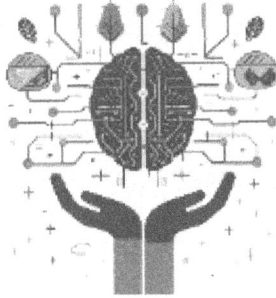

THE NONPROFIT AI PLAYBOOK

Your Essential Guide to Implementing AI for Marketing, Fundraising, Operations, and Impact

BY

DR. VICTORIA BOYD

PRODUCED BY

THE PHILANTREPRENEUR FOUNDATION

THE NONPROFIT AI PLAYBOOK
Your Essential Guide to Implementing AI for Marketing, Fundraising, Operations, and Impact

Copyright © 2025 The Philantrepreneur Foundation aka Dr. Victoria Boyd

Bulk orders are available, Contact:
marketing@philantrepreneurfoundation.org

ISBN: 978-0-9854219-4-6 – ASIN: 0985421940
eBook ISBN: 978-0-9854219-5-3 - **ASIN**: B0F2SLVMJ4
Categories:
Nonfiction > Business & Money > Business Development Entrepreneurship >Nonprofit Organizations & Charities
Nonfiction > Computer & Technology > Business Technology
Nonfiction >Business & Money > Management & Leadership > Planning & Forecasting

The Philantrepreneur Foundation
304 S. Jones Ste. 3084
Las Vegas, Nevada

TABLE OF CONTENTS

FOREWORD

In today's rapidly evolving landscape, if you haven't already started using AI, you are already behind. The integration of artificial intelligence into nonprofit operations is no longer a luxury; it is a necessity for organizations striving to maximize their impact and remain relevant in a competitive environment.

As the author of this playbook, my journey into the world of AI has been anything but conventional. I do not consider myself a digital developer or a tech geek; rather, I am a passionate writer who first explored AI as a tool for generating topics and ideas. Initially, I relied on simple bullet point lists, but my curiosity soon led me to ask deeper questions rather than merely seeking results. This shift in perspective opened my eyes to the vast potential of AI.

Driven by my inquisitiveness, I enrolled in a course on crafting effective prompts, which transformed my understanding of how to engage with AI tools. I realized that my knowledge barely scratched the surface, and I was only utilizing a small component of the generative capabilities available. To deepen my understanding, I pursued several additional courses from various perspectives on AI, each offering valuable insights into its applications.

However, it wasn't until I completed a certification course that I truly connected the dots. This course provided me with a global perspective on AI, reinforcing the importance of ethical and responsible use in the nonprofit sector. I became acutely aware of the challenges and opportunities that AI presents for organizations dedicated to social good.

I understand that many of you may have fears about adopting new technologies. Change can be daunting, and stepping out of your comfort zone is never easy. But I urge you to embrace this change, just as I did. Taking the leap into AI can be one of the most valuable decisions for your organization. By doing so, you will not only enhance your operational efficiency but also empower your mission to create meaningful change in your community.

As you delve into this playbook, I encourage you to view AI as a powerful ally in your mission. Understanding and implementing AI is not just about keeping up with technology; it is about enhancing your ability to create lasting impact. Together, we can harness the power of AI to drive innovation, improve efficiency, and ultimately amplify our impact.

Dr. Victoria Boyd, President
The Philantrepreneur Foundation

PART ONE: UNDERSTANDING AI AND ITS POTENTIAL

Chapter 1

The Future is Here: AI is Essential for Nonprofits to Survive and Thrive

Nonprofits are the heart and soul of social change, tackling some of the toughest problems out there. From fighting hunger to standing up for human rights, these organizations are all about making a real difference, often with limited funds, small teams, and super-tight budgets. In this situation, every choice, every tool, and every plan has to be spot-on to get the most out of their efforts.

Artificial intelligence (AI) isn't some far-off idea anymore; it's something nonprofits need right now. Those that don't jump on the AI bandwagon risk falling behind, struggling to keep up with rising needs and a world that's increasingly digital. AI isn't just an option; it's key to surviving and doing well in today's nonprofit world. Whether it's making workflows smoother, helping with better decision-making, or reaching more people, AI is becoming a must-have for organizations with a mission.

But AI isn't just about robots or crunching numbers; it's about giving nonprofits the power to do more with less. This book isn't just about getting what AI is; it's about learning how to use it to make real change happen. Whether you're a small, local group trying to get the word out or a larger nonprofit looking to grow, "The Nonprofit AI Playbook: Your Essential Guide to Implementing AI for Marketing, Fundraising, Operations, and Impact" will give you the tools and plans you need to get started with AI now, not later.

The Clock is Ticking: Why Nonprofits Need to Act Fast

The nonprofit world is at a turning point. The need for services is growing, donors expect more, and everything is moving faster online. Organizations that don't add AI to their game plan will struggle to stay in the game and keep going. The challenges are clear:

- Limited Funds – Many nonprofits run on tight budgets, making it tough to invest in new tools or tech.
- Time Crunch – With small teams juggling a million things, finding time to try new things can be a real struggle.
- Staffing Shortages – With few people, organizations have to do more with less, often relying on volunteers and part-time help.
- Getting Noticed – Having a great mission isn't enough; nonprofits have to connect with people, attract donors, and share their stories in a way that resonates.

Nonprofits can't afford to wait. The organizations that use AI-powered plans today will not just survive but really take off, becoming more efficient, getting donors more involved, and making a bigger impact than they ever thought possible.

AI: A Game-Changer for Nonprofits

AI can totally change how nonprofits work, and those that embrace it will have an advantage in achieving their goals. With AI, nonprofits can:

- Work More Efficiently – Automate tasks like entering data, managing donors, and answering emails, freeing up staff for important work.
- Make Smarter Decisions – Use AI analytics to spot trends, predict fundraising results, and make choices based on solid data.
- Reach More People – Personalize communications with donors, target the right audiences, and improve marketing to get more people involved.
- Tell Better Stories – Use AI to create compelling stories, show data in a clear way, and make reporting easier for stakeholders and funders.

A Simple Guide to Using AI

AI isn't here to replace the human connection that's so important in nonprofit work; it's here to make the efforts of dedicated teams even better. When used right, AI can multiply your efforts, helping organizations reach more people, increase their impact, and work more strategically.

Getting started with AI might seem like a lot, especially if you're not familiar with the tech. This book is a simple guide, walking nonprofits through the steps of adding AI. It starts with getting what AI is, how it works, and why it's important for nonprofits. From there, we look at how AI can be used in marketing, fundraising, operations, planning, and measuring impact.

Each chapter is made to give you practical advice, showing nonprofits how to use AI responsibly and effectively. We'll also clear up common myths and misunderstandings, making sure you have a realistic view of what AI can and can't do.

At the end of each chapter, you'll find a reflection section with a simple question. Take time to think about it as we try to help you take steps to use AI right away.

Why This Book is Important

Using AI isn't optional anymore; it's becoming necessary for organizations that want to stay effective and competitive. Nonprofits have typically been behind the corporate world in using technology, but that's changing. Organizations that take a forward-thinking approach to AI will be better prepared for the future.

We'll simplify AI and give nonprofit pros a practical guide to add AI to their work. Whether you're an executive director, fundraiser, program manager, or marketing specialist, this book will help you:

- Grasp AI ideas and how they relate to nonprofits
- Find the right AI tools for your organization

- Use AI-driven plans for marketing, fundraising, and operations
- Avoid common mistakes and ethical issues in AI use
- Measure and share the impact of AI-enhanced programs

This isn't a book about AI theory, it's about putting AI to work for your mission. By the end of this book, you'll not just understand AI basics, but you'll also have a clear plan to add AI to your nonprofit's strategy. To help you take action right away, we've included questions at the end of most chapters. These questions are made to:

1. Get you thinking about how AI can address your organization's specific challenges.
2. Help you develop steps that are specific to your nonprofit.
3. Help you really engage with the material, making sure you understand key ideas and use them effectively.
4. Support ongoing learning by giving you a place to document your ideas, progress, and next steps.

Let's Get Started

AI is changing the future of nonprofit work, and the possibilities are endless. Whether you're just starting to explore AI or looking to improve your current plans, "The Nonprofit AI Playbook: Your Essential Guide to Implementing AI for Marketing, Fundraising, Operations, and Impact" will be your guide.

AI is here, and nonprofits need to act now to stay relevant, get the most out of their resources, and increase their impact. Those that embrace AI will survive and thrive—those that ignore it risk being left behind. Let's start this journey together.

Question: What are the top 3 challenges your nonprofit faces that AI could help address?

Write down your initial thoughts on how AI could support your mission.

_____

_____

_____

_____

_____

_____

_____

_____

_____

_____

_____

_____

_____

_____

_____

_____

_____

_____

_____

_____

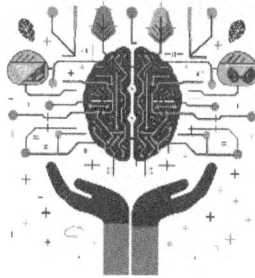

Chapter 2
Busting Myths
Clarifying Misconceptions and Avoiding Pitfalls

Artificial Intelligence (AI) is one of the most exciting and transformative technologies of our time, but it's also one of the most misunderstood. For nonprofits, AI offers incredible opportunities to streamline operations, enhance donor engagement, and amplify impact. However, many organizations approach AI with unrealistic expectations or fall into common traps that hinder their success.

This chapter will **debunk the most pervasive myths about AI** and provide **practical advice** to help nonprofits avoid costly mistakes. By understanding what AI can and cannot do, you'll be better equipped to leverage it as a powerful tool for your mission. Let's dive into the myths and set the record straight.

Myth 1: AI is a Magic Solution That Solves All Problems Instantly

It's easy to think of AI as a magic wand, something you can wave to instantly fix all your organization's challenges. But the truth is, AI is a tool, not a miracle worker. While it can automate tasks, analyze data, and provide insights, it requires **thoughtful implementation** and **clear goals** to be effective.

Think of AI like a new team member. You wouldn't hire someone and expect them to know exactly what to do without any guidance, right? The same goes for AI. Your AI tool must be "taught" what goals and outcomes match your needs. For example, if you're using AI to personalize donor communications, you'll need to provide it with data about your donors' preferences and behaviors.

It's also important to remember that AI is not meant to replace staff. Instead, it's designed to **enhance their performance**. By automating repetitive tasks like data entry or email responses, AI frees up your team's time to focus on what really matters, your mission.

Myth 2: AI Doesn't Require Clear Objectives or Goals

One of the biggest mistakes nonprofits make is jumping into AI without a clear plan. They assume that simply having AI tools in place will automatically lead to better outcomes. But the reality is, AI is most effective when it's aligned with your organization's goals. Without clear objectives, you risk wasting time and resources on tools that don't deliver meaningful results.

For example, if you're using AI for marketing, you'll need to define what success looks like. Is it increasing donor engagement? Expanding your reach on social media? Once you've set your goals, you can use AI tools like **Galaxy.ai**, and resources in **Nonprofit AI Playbook Toolkit** to achieve them.

A key part of this process is creating a **brand profile**. This is essentially a set of guidelines that tells your AI tool how to represent your organization, your tone of voice, key messaging, and visual style. By teaching your AI tool how to "speak" like your nonprofit, you can ensure consistency across all your communications.

Another important skill is **learning how to prompt** your AI tool effectively. This means giving it clear, specific instructions to get the results you want. For example, instead of asking, "Write a social media post," you could say, "Write a social media post about our upcoming gala, using a friendly and enthusiastic tone."

Finally, always **check for accuracy**. AI is powerful, but it's not perfect. Make sure to review its outputs carefully to ensure they align with your goals and values.

Myth 3: AI Can Work with Poor-Quality Data

You've probably heard the phrase "garbage in, garbage out." This couldn't be truer for AI. Many nonprofits assume that AI can produce accurate insights and predictions even if the data it's trained on is incomplete, outdated, or biased. But the reality is, AI is only as good as the data it's given.

Poor-quality data leads to inaccurate results, flawed decisions, and potential harm to your organization's reputation. For example, if your donor data is outdated, your AI tool might send personalized messages to people who are no longer interested in your cause.

To avoid this, start by **cleaning your data**. Remove duplicates, correct errors, and ensure your data is up to date. It's also important to use **diverse data** that represents the full range of your stakeholders. This helps prevent bias and ensures your AI tool works for everyone.

Finally, **regularly audit your data**. AI is not a "set it and forget it" solution. You'll need to continuously monitor and improve the quality of your data inputs to get the best results.

Myth 4: AI Tools Are Plug-and-Play with No Learning Curve

There's a common misconception that AI tools are easy to use and require no training or expertise. But the reality is, AI tools often come with a **learning curve**. While platforms like **Galaxy.ai** and the **AI Playbook Toolkit** are designed to be user-friendly, they still require some effort to master.

For example, you'll need to learn how to navigate the platform, input your data, and interpret the results. You might also need to experiment with different settings to get the outcomes you want.

The good news is this learning curve is manageable. Start with simple tasks, like automating email responses or generating social media posts. As you become more comfortable with the tool, you can explore more advanced features, like predictive analytics or donor segmentation.

Myth 5: AI Can Replace Human Creativity and Decision-Making

AI is incredibly powerful, but it's not a substitute for human creativity and decision-making. It's AI can analyze data and generate insights, it's still up to your team to interpret those insights and make strategic decisions.

For example, AI can help you identify trends in donor behavior, but it's your team's job to decide how to act on those trends. Similarly, AI can generate content, but it's your team's creativity that brings that content to life.

The key is to view AI as a **collaborator**, not a replacement. By combining the strengths of AI with the creativity and expertise of your team, you can achieve even greater impact.

Additional Warnings and Myths
Myth 6: AI is Too Expensive for Nonprofits

While AI tools can be costly individually, there are many affordable options available, like **Galaxy.ai** offering multiple tools at a fraction of the cost.

Myth 7: AI is Only for Tech-Savvy Organizations

You don't need to be a tech expert to use AI. Many tools are designed with nonprofits in mind, offering intuitive interfaces and step-by-step guides to help you get started.

Myth 8: AI is a One-Time Investment

AI is not a one-and-done solution. It requires ongoing maintenance, updates, and training to stay effective. Make sure to budget for these costs and allocate time for your team to learn and adapt.

Practical Tips for Avoiding Mistakes

1. **Start Small:** Focus on one specific problem AI can solve, like automating donor emails or analyzing survey data.
2. **Set Clear Goals:** Define what you want to achieve with AI and how you'll measure success.
3. **Invest in Training:** Provide your team with the resources and support they need to learn and adapt to AI tools.
4. **Monitor and Adjust:** Regularly review your AI tools and processes to ensure they're delivering the results you want.

By debunking these myths and following these tips, you'll be well on your way to successfully integrating AI into your nonprofit's operations. Remember, AI is a tool, and like any tool, it's most effective when used thoughtfully and strategically.

Question: Which AI myth resonated with you, and how will you address it in your organization?

Outline a plan to avoid common AI pitfalls in your nonprofit.

_____

_____

_____

_____

_____

_____

_____

_____

_____

_____

_____

_____

_____

_____

_____

_____

_____

_____

_____

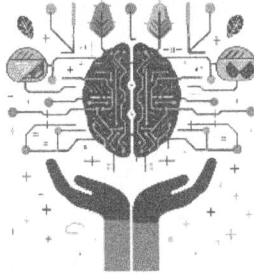

Chapter 3
AI Fundamentals for Nonprofits

Artificial Intelligence (AI) is transforming the nonprofit sector, providing organizations with powerful tools to enhance efficiency, improve decision-making, and amplify impact. However, for many nonprofits, AI remains an unfamiliar and sometimes intimidating concept. The goal of this chapter is to build a strong foundation in AI, ensuring that nonprofit professionals understand its core principles, relevant tools, and practical applications. By the end of this chapter, you'll have the knowledge to confidently explore AI solutions that align with your mission.

What is AI, and How Does It Work?
AI refers to the simulation of human intelligence in machines. These systems are designed to think, learn, and make decisions, often

performing tasks that would typically require human intervention. AI can analyze vast amounts of data quickly and accurately, recognize patterns, understand language, and solve complex problems.

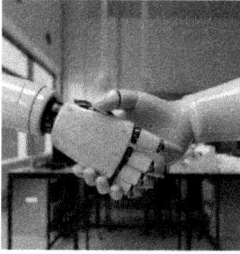

AI Operates Through a Collaboration of Technologies

AI is not a single technology but a collection of interrelated technologies that work together to achieve intelligent behavior. Below are the core technologies that power AI systems:

- *Machine Learning (ML):* Machine learning is a method where computers are trained to recognize patterns in data and make predictions or decisions without being explicitly programmed. Instead of following a fixed set of rules, ML models improve their accuracy over time by analyzing new data. ML is commonly used in predictive analytics, recommendation systems, and fraud detection. For example, a nonprofit might use ML to predict donor behavior by analyzing past contributions and engagement levels.

- *Deep Learning*: Deep learning is an advanced form of machine learning that mimics the way the human brain processes information. It uses artificial neural networks with multiple layers (hence "deep") to analyze and interpret complex patterns in large datasets. Each layer of the network processes specific features of the data, gradually refining the understanding of the input. Deep learning is used for tasks that require advanced pattern recognition, such as image recognition, speech processing, and autonomous systems. For example, a deep learning model can analyze thousands of grant applications and identify common success factors.

- *Natural Language Processing* (NLP): NLP enables computers to understand, interpret, and generate human language in a way that is both meaningful and contextually relevant. It allows AI systems to process text and speech, making them useful for applications like chatbots, sentiment analysis, and real-time translations. In a nonprofit setting, NLP can be used to analyze donor feedback from surveys, automate responses in a chatbot, or translate materials for multilingual outreach.

- *Generative AI:* Generative AI is a branch of artificial intelligence focused on creating new content, such as text, images, music, or videos. It works by learning from vast datasets and then generating outputs based on patterns it has recognized. This technology is behind tools like ChatGPT, which can draft grant proposals or fundraising emails, and AI-powered design tools that generate social media graphics. Nonprofits can leverage generative AI to automate content creation, saving time and ensuring consistent messaging.

Each of these technologies plays a crucial role in how AI functions and provides practical benefits to nonprofits looking to improve efficiency and effectiveness in their operations.

> **The FULL AI Glossary is available in the Nonprofit AI Playbook Toolkit.**

Types of AI

1. Narrow AI (Weak AI)

Narrow AI, also known as Weak AI, is designed to perform specific tasks within a limited domain. It does not possess human-like intelligence but excels at predefined functions. Some real-world examples include:

- Voice Assistants: AI-powered assistants like Siri, Alexa, and Google Assistant, which respond to voice commands and perform tasks like setting reminders or answering questions.
- Recommendation Engines: Platforms like Netflix and Amazon use AI to analyze user preferences and suggest movies, books, or products based on past behavior.
- Spam Filters: Email services like Gmail utilize AI to detect spam messages and keep inboxes free from unwanted content.
- Chatbots: Many nonprofits use AI chatbots to provide automated responses to donor inquiries or assist with volunteer applications.

2. General AI (Strong AI)

General AI, or Strong AI, is a theoretical form of AI that has not yet been achieved. Unlike Narrow AI, General AI would have human-like cognitive abilities, enabling it to understand, learn, and apply knowledge across a wide range of activities without specific programming. If developed, General AI could:

- Think critically and make decisions independently.
- Adapt to new environments and solve unfamiliar problems.
- Exhibit creativity and emotional intelligence akin to human interactions. Currently, all AI in existence is classified as Narrow

AI, as General AI remains a goal for future advancements in AI research.

3. Machine Learning (ML)

Machine Learning (ML) is a subset of AI that enables machines to improve their performance over time by learning from data rather than following explicit programming. ML models identify patterns in large datasets and use these insights to make predictions or automate decision-making. Examples include:

- Fraud Detection: Banks and financial institutions use ML to detect suspicious transactions and prevent fraud.
- Donor Engagement Prediction: Nonprofits use ML to analyze donor behavior and predict which individuals are most likely to contribute again.
- Personalized Marketing: AI-powered email marketing tools analyze open rates and engagement metrics to tailor content for each recipient.

Types of Machine Learning:

1. Supervised Learning: The algorithm is trained using labeled data (e.g., a dataset of past donor behaviors with corresponding donation outcomes).
2. Unsupervised Learning: The algorithm identifies patterns and structures in unlabeled data (e.g., grouping donors with similar giving patterns without predefined categories).

3. Reinforcement Learning: The AI learns through trial and error by receiving rewards or penalties for actions (e.g., an AI-powered chatbot refining its responses based on user interactions).

4. Deep Learning

Deep Learning is an advanced subset of Machine Learning that utilizes artificial neural networks with multiple layers (hence "deep") to analyze complex patterns in large datasets. Deep Learning is especially effective for:

- Image Recognition: Used in nonprofit advocacy campaigns to analyze social media images and detect visual trends.
- Language Processing: AI-driven translation tools like Google Translate rely on deep learning to interpret and convert text between languages.
- Speech Recognition: Virtual assistants and transcription services use deep learning to understand spoken language.

Deep Learning is particularly valuable for tasks requiring a high level of pattern recognition, such as predicting donor behavior or analyzing vast amounts of program impact data.

> Download the FULL *Nonprofit AI Playbook Glossary* - as a book owner you have full access. See *Resources* for details.

AI Bot

Now that we've explored the fundamentals of AI and some of its applications, let's dive into a term that's generating significant buzz in the market, the **AI Bot.** You may have heard it referred to as an **AI Employee**, **AI Assistant**, or **AI Agent**, but what exactly is it, and how can it benefit your nonprofit? Let's clarify its capabilities and limitations, and show you how it can become a valuable asset in your organization's toolkit.

Definition and Description

An AI bot (also referred to as an AI Employee, AI Assistant, or AI Agent) is a software application powered by artificial intelligence (AI) that performs tasks autonomously or semi-autonomously, often mimicking human-like interactions and decision-making processes.

These bots are designed to assist individuals, teams, or organizations by automating repetitive tasks, analyzing data, and providing insights, thereby enhancing efficiency and productivity.

AI bots leverage technologies such as natural language processing (NLP), machine learning (ML), and generative AI to understand, interpret, and respond to user inputs. They can be integrated into various platforms, including websites, messaging apps, and enterprise software, to provide seamless support and functionality.

Key Features of an AI Bot

1. Natural Language Understanding (NLU): Ability to interpret and respond to human language in a conversational manner.
2. Task Automation: Capability to perform specific tasks without human intervention.
3. Data Integration: Ability to connect with databases, CRMs, and other systems to access and analyze information.
4. Scalability: Can handle multiple tasks or interactions simultaneously, making it ideal for large-scale operations.

5. Customization: Can be tailored to meet the specific needs of an organization or individual.

What an AI Bot Can Do

1. Automate Repetitive Tasks
   - Perform routine administrative tasks such as scheduling, data entry, and email management.
   - Example: An AI bot can schedule meetings, send reminders, and update calendars.

2. Provide Customer Support
   - Answer frequently asked questions, resolve common issues, and guide users through processes.
   - Example: A chatbot on a nonprofit website can assist donors with donation inquiries.

3. Analyze Data and Generate Insights
   - Process large datasets to identify trends, patterns, and actionable insights.
   - Example: An AI bot can analyze donor data to predict giving behavior and recommend engagement strategies.

4. Assist with Content Creation
   - Generate text, images, or videos for marketing, fundraising, or reporting purposes.
   - Example: An AI bot can draft social media posts, grant proposals, or impact reports.

5. Enhance Decision-Making
   - Provide recommendations based on data analysis and predefined criteria.

- Example: An AI bot can suggest optimal times to launch fundraising campaigns based on historical data.

6. Facilitate Communication
   - Translate languages, summarize conversations, and provide real-time responses.
   - Example: An AI bot can translate donor communications into multiple languages for global outreach.

7. Streamline Workflows
   - Integrate with existing systems to automate workflows and improve efficiency.
   - Example: An AI bot can manage volunteer sign-ups and send automated onboarding emails.

What an AI Bot Cannot Do

1. Replace Human Creativity and Empathy
   - While AI bots can generate content and provide recommendations, they lack the ability to think creatively or empathize with human emotions.
   - Example: An AI bot cannot craft a deeply personal story that resonates emotionally with donors.

2. Make Ethical or Moral Judgments
   - AI bots operate based on data and algorithms, not ethical principles. They cannot make moral decisions or understand the nuances of ethical dilemmas.
   - Example: An AI bot cannot determine whether a fundraising strategy aligns with your nonprofit's ethical values.

3. Handle Highly Complex or Unpredictable Situations

- AI bots are limited to the tasks and scenarios they are programmed for. They struggle with ambiguity or situations requiring human intuition.
- Example: An AI bot cannot mediate a conflict between team members or adapt to unexpected crises.

4. Learn Without Data
   - AI bots rely on data to learn and improve. Without sufficient or high-quality data, their performance may be limited.
   - Example: An AI bot cannot predict donor behavior if it lacks historical donor data.

5. Operate Independently Without Oversight
   - AI bots require human supervision to ensure accuracy, relevance, and ethical use. They cannot fully replace human judgment or accountability.
   - Example: An AI bot's outputs must be reviewed and validated by a human before implementation.

Examples of AI Bots in Action

1. Fundraising Assistant: Automates donor outreach, tracks engagement, and suggests personalized communication strategies.
2. Grant Writing Assistant: Helps identify funding opportunities, drafts grant proposals and ensures compliance with funder guidelines.

3. Volunteer Coordinator: Manages volunteer sign-ups, schedules shifts, and sends reminders.

4. Marketing Assistant: Generates social media content, analyzes campaign performance, and recommends optimizations.
5. Customer Support Bot: Provides 24/7 assistance to donors, volunteers, and beneficiaries through chat or email.

Benefits of Using an AI Bot

1. Increased Efficiency: Automates time-consuming tasks, freeing up staff to focus on high-impact work.
2. Cost Savings: Reduces the need for additional personnel or resources to handle repetitive tasks.
3. Improved Accuracy: Minimizes human error in data entry, analysis, and reporting.
4. Enhanced Engagement: Provides instant, personalized responses to stakeholders, improving their experience.
5. Scalability: Handles large volumes of tasks or interactions without compromising quality.

Limitations and Ethical Considerations

1. Bias in Data: AI bots may perpetuate biases present in the data they are trained on.
2. Privacy Concerns: Handling sensitive data requires robust security measures to protect user privacy.
3. Dependence on Technology: Over-reliance on AI bots can lead to reduced human oversight and accountability.
4. Transparency: Organizations must be transparent about the use of AI bots to maintain trust with stakeholders.

An AI bot is a powerful tool that can transform how nonprofits operate, enabling them to work smarter, faster, and more effectively. However, it is not a replacement for human creativity, empathy, or judgment. By understanding what an AI bot can and cannot do, nonprofits can leverage this technology responsibly to achieve their missions and create lasting impact.

AI Applications in Nonprofits AI helps nonprofits work more efficiently and with greater insight, supporting the organization's broader strategy rather than replacing it. Here are key areas where AI can make a significant impact:

1. **Donor Engagement & Fundraising**
   - Predictive Analytics: AI analyzes donor data to predict who is most likely to give, how much, and when. This helps nonprofits focus their efforts on high-potential donors.
   - Personalized Communication: AI-powered chatbots or email systems can personalize outreach to engage donors at scale, improving retention and donations.
   - Automating Donation Processing: AI systems can streamline online donation processing, reducing manual errors and improving the donor experience.

2. **Grant Writing & Research**
   - Automated Research Tools: AI can search for relevant grants or funding opportunities, saving hours of manual work.

- Grant Writing Assistance: AI tools can help draft proposals, especially in structuring language and ensuring it meets funder criteria.

3. **Program Management**
   - Data Analysis for Impact: AI helps track outcomes and analyze data from program participants, improving decision-making and demonstrating program effectiveness to stakeholders.
   - Automation of Administrative Tasks: AI can automate routine tasks, like scheduling or reporting, allowing staff to focus on higher-value work.

4. **Volunteer Recruitment & Management**
   - Matching Algorithms: AI can match volunteers with roles that align with their skills and availability, improving volunteer engagement and retention.
   - AI-Powered Onboarding: Chatbots can guide volunteers through the onboarding process, providing 24/7 support.

5. **Marketing & Awareness**
   - Content Creation: AI tools can generate social media posts, blog content, or even marketing campaigns, freeing up time for staff to focus on strategy.
   - Audience Segmentation: AI analyzes data to segment audiences for targeted outreach, ensuring that messaging reaches the right people with personalized content.

6. **Fraud Detection and Risk Management**
   - AI Monitoring: AI systems can detect irregularities in financial transactions or program data, helping to identify potential fraud or areas of risk before they escalate.

AI is no longer a futuristic concept; it is a present-day necessity for nonprofits aiming to increase their impact. By understanding AI fundamentals, leveraging key tools, and integrating AI-powered solutions into your workflow, your nonprofit can operate more efficiently and effectively.

Question: Which area of your nonprofit's operations (e.g., fundraising, marketing, operations) could benefit most from AI?

Note Space: Identify 1-2 AI tools or strategies you'd like to explore further.

_____

_____

_____

_____

_____

_____

_____

_____

_____

_____

_____

_____

_____

_____

_____

_____

_____

_____

_____

_____

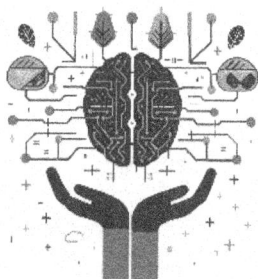

Part 2: AI in Action – Practical Applications

Chapter 4

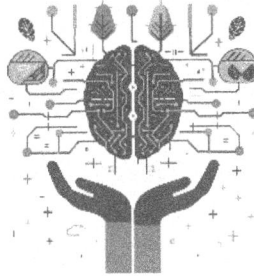

Nonprofit AI Playbook's
AI Responsibilities Framework

As we embark on the journey of integrating artificial intelligence (AI) into nonprofit operations, it is essential to establish a solid foundation of ethical and responsible practices. Before diving into the specific applications of AI, such as fundraising, program delivery, and operations, this chapter introduces the **AI Responsibilities Framework**.

This framework is designed to guide nonprofits in harnessing the power of AI while upholding their core values and commitments to their communities. It emphasizes the importance of ethical oversight, inclusivity, and accountability in all AI initiatives. By understanding and adhering to these key responsibilities, organizations can ensure that their use of AI not only enhances their

effectiveness but also aligns with their mission to create positive social change.

In this chapter, we will explore the fundamental principles that underpin responsible AI use, including governance, transparency, and equity. We will also provide actionable steps and best practices that nonprofits can implement to navigate the complexities of AI adoption. By equipping yourself with this knowledge, you will be better prepared to leverage AI tools effectively and ethically, setting the stage for successful applications in the chapters that follow.

Let's delve into the **AI Responsibilities Framework** and discover how it can empower your organization to use AI responsibly and effectively in pursuit of your mission.

What Sets This Framework Apart?

1. **Inclusivity**: This framework goes beyond fundraising to address all aspects of nonprofit operations—governance, program delivery, human resources, financial management, and more. It ensures AI is applied equitably and benefits all stakeholders, including marginalized communities.

2. **Alignment with Global AI Ethics Standards**: Drawing on principles from leading global organizations, including the **Partnership on AI (PAI)**, the **IEEE Global Initiative on Ethics of Autonomous and Intelligent Systems**, the **European Commission's High-Level Expert Group on AI (AI HLEG)**, the **Montreal Declaration for Responsible AI**, the **OECD Principles on AI**, and the **United**

Nations' AI for Good Initiative, this framework emphasizes fairness, transparency, accountability, and privacy. It ensures your organization aligns with global standards while addressing local needs.

3. **Nonprofit Development Best Practices**: Built on decades of nonprofit expertise, this framework incorporates proven strategies for systems and operations. It focuses on scalability, sustainability, and mission alignment, helping you maximize impact while minimizing risks.

4. **Systems and Operations Focus**: Recognizing that nonprofits operate in complex environments, this framework takes a holistic approach. It ensures AI tools integrate seamlessly into your workflows, enhance efficiency, and support long-term organizational growth.

Why This Framework Matters for Nonprofits

Nonprofits play a critical role in addressing societal challenges, and AI has the potential to amplify their impact. However, without a clear and ethical framework, AI can inadvertently perpetuate biases, create inefficiencies, or undermine public trust. This playbook provides the tools and guidance you need to:

- **Responsibly leverage AI** to enhance your programs, operations, and fundraising efforts.
- **Build trust** with donors, beneficiaries, and stakeholders by demonstrating ethical AI use.
- **Future-proof your organization** by adopting scalable and sustainable AI practices.

Let's build a future where technology serves humanity, one nonprofit at a time.

A Case-Use Example: Ethics in Action

Consider a mid-sized nonprofit focused on education equity that decided to adopt AI tools to enhance its programs. Here's how they used the **Nonprofit AI Playbook's AI Responsibilities Framework** to ensure ethical and responsible implementation:

1. **Ethical Oversight and Transparency**
   The nonprofit established an AI ethics committee, including staff, board members, and community representatives, to oversee AI use. They developed clear policies on how AI decisions would be made and communicated these to stakeholders, ensuring transparency and accountability.

2. **Equity and Bias Mitigation**
   When implementing an AI tool to analyze student performance data, the nonprofit used the framework's equity guidelines to identify and address potential biases. They ensured the tool was trained on diverse datasets and regularly audited its outputs to prevent unfair outcomes for marginalized students.

3. **Privacy and Data Protection**
   To safeguard student data, the nonprofit followed the framework's privacy principles. They implemented robust data security measures, obtained informed consent from families, and ensured compliance with data protection laws like GDPR and FERPA.

4. **Mission Alignment and Impact Assessment**
   Before adopting AI, the nonprofit used the framework to assess

whether the tools aligned with their mission of education equity. They conducted a pilot program to evaluate the tool's impact on student outcomes and adjusted their approach based on feedback from teachers and students.

5. **Stakeholder Engagement and Trust Building**
   The nonprofit engaged parents, teachers, and students in the decision-making process, using the framework's community engagement guidelines. By involving stakeholders early and often, they built trust and ensured the AI tools met the community's needs.

6. **Continuous Monitoring and Improvement**
   After implementation, the nonprofit established feedback loops to monitor the AI tool's performance and ethical implications. They used the framework's evaluation metrics to identify areas for improvement and adapted their practices accordingly.

By following the **Nonprofit AI Playbook's AI Responsibilities Framework,** this nonprofit was able to harness the power of AI while avoiding ethical pitfalls, building trust with stakeholders, and staying true to their mission of education equity.

This document is available for download
in the Nonprofit AI Playbook Toolkit.

NONPROFIT AI RESPONSIBILITIES FRAMEWORK

Key Principles for AI Use Across All Areas

- Ethical Use: Prioritize fairness, transparency, and accountability in all AI applications.
- Inclusivity: Ensure AI tools are accessible and beneficial to all stakeholders, including marginalized groups.
- Sustainability: Use AI to support long-term organizational and community well-being.
- Collaboration: Engage stakeholders in the development and implementation of AI initiatives.

1. Governance and Leadership

- Ethical Oversight: Establish a board or committee to oversee AI use, ensuring alignment with the organization's mission and values.
- Transparency: Develop clear policies on how AI decisions are made and communicated to stakeholders.
- Accountability: Define roles and responsibilities for AI implementation and monitoring.
- Strategic Alignment: Ensure AI initiatives support the organization's long-term goals and mission.

2. Fundraising

- Donor Engagement: Use AI to personalize donor outreach and improve relationship management.
- Predictive Analytics: Leverage AI to identify potential donors and forecast fundraising outcomes.
- Campaign Optimization: Implement AI to analyze and optimize fundraising campaigns for maximum impact.

- Ethical Fundraising: Ensure AI tools are used transparently and ethically, avoiding manipulation or bias.

3. Program and Service Delivery

- Impact Assessment: Use AI to evaluate the effectiveness of programs and services, ensuring they meet community needs.
- Personalization: Leverage AI to tailor services to individual beneficiaries, improving outcomes.
- Equity: Ensure AI tools do not perpetuate biases and are accessible to all the populations served.
- Resource Allocation: Use AI to optimize the distribution of resources for programs and services.

4. Operations and Administration

- Efficiency: Implement AI to streamline administrative tasks such as scheduling, reporting, and resource allocation.
- Data Management: Use AI to enhance data security, accuracy, and accessibility across the organization.
- Risk Management: Identify and mitigate risks associated with AI use in operations.
- Process Automation: Automate repetitive tasks to free up staff time for higher-value activities.

5. Human Resources

- Recruitment and Retention: Use AI to improve hiring processes, ensuring diversity and inclusion.
- Training and Development: Provide staff with training on AI tools and ethical considerations.

- Performance Management: Implement AI to support fair and objective performance evaluations.
- Employee Engagement: Use AI to gather feedback and improve workplace satisfaction.

6. Financial Management

- Budgeting and Forecasting: Use AI to improve financial planning and resource allocation.
- Fraud Detection: Implement AI to identify and prevent financial irregularities.
- Donor Management: Expand AI use to manage donor relationships, ensuring ethical and effective engagement.
- Expense Tracking: Use AI to monitor and optimize organizational spending.

7. Technology and Infrastructure

- System Integration: Ensure AI tools are compatible with existing systems and workflows.
- Cybersecurity: Use AI to enhance protection against cyber threats.
- Scalability: Plan for AI systems that can grow with the organization's needs.
- Maintenance and Support: Establish protocols for maintaining and updating AI systems.

8. Community and Stakeholder Engagement

- Feedback Mechanisms: Use AI to gather and analyze feedback from beneficiaries, donors, and partners.

- Communication: Implement AI to improve outreach and engagement strategies.
- Collaboration: Foster partnerships with other organizations to share AI best practices and resources.
- Public Trust: Ensure AI use aligns with community expectations and builds trust.

9. Compliance and Legal

- Regulatory Adherence: Ensure AI use complies with all relevant laws and regulations.
- Privacy Protection: Implement AI tools that safeguard personal and sensitive information.
- Ethical Standards: Develop and enforce ethical guidelines for AI use across the organization.
- Data Governance: Establish policies for data collection, storage, and usage.

10. Evaluation and Continuous Improvement

- Performance Metrics: Establish metrics to evaluate the impact of AI on organizational goals.
- Feedback Loops: Create mechanisms for ongoing feedback and improvement of AI systems.
- Innovation: Encourage a culture of innovation, exploring new AI applications to enhance operations.
- Learning and Adaptation: Use AI insights to adapt strategies and improve outcomes over time.

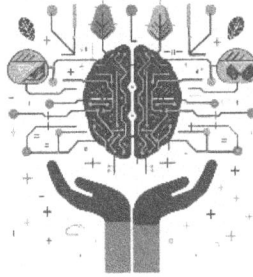

Chapter 5

The TPF PROMPT Method:
Crafting Effective Prompts for AI

The Power of Prompts in AI for Nonprofits

In today's fast-paced digital world, artificial intelligence (AI) has become an indispensable tool for nonprofits looking to amplify their impact. At the core of this technology lies the **prompt**, a set of specific instructions that guides AI systems to generate the content, insights, or solutions you need.

Have you ever heard of a 'brain dump" where you dump everything out, concepts, ideas, strategies, to resort it later. Think of a prompt as the bridge between your organization's goals (the brain dump) and the AI's capabilities to organize and

develop the desired results. The better the prompt, the better the results.

For nonprofits, mastering the art of prompting is not just a technical skill, it's a strategic advantage. With limited resources and time, organizations must ensure that every interaction with AI tools is purposeful and aligned with their mission. Whether you're drafting a fundraising email, analyzing donor data, or creating social media content, a well-crafted prompt can unlock AI's full potential, delivering **insightful, relevant, and actionable outcomes**.

Consider this example: A nonprofit using an AI content generator put the prompt, *"Create a social media post highlighting our upcoming fundraising event and its impact on the community."* It may sound reasonable but was it enough to give the tool data it needed? By providing clear instructions, the organization ensures the AI produces content that resonates with its audience and aligns with its goals. Clear instructions include setting up a brand profile that becomes your foundational information.

However, crafting effective prompts requires more than just typing a sentence. It's about **being specific, intentional, and iterative**. Nonprofits must define the desired tone, format, and audience, and be prepared to refine their prompts based on the AI's responses. This process of experimentation and refinement is key to maximizing the value of AI tools.

By learning how to develop high-quality prompts, nonprofits can **streamline operations, enhance donor engagement, and amplify their storytelling efforts**.

TPF PROMPT METHOD

Introducing the **TPF PROMPT METHOD**: a structured approach designed specifically for nonprofits to craft effective prompts when using AI tools. This method empowers organizations to enhance the quality of the content generated by AI, ensuring that it aligns with their mission and meets the needs of their audience. By following the **TPF PROMPT METHOD**, nonprofits can confidently navigate the world of AI, making greater impact in their communities and achieving their goals more efficiently.

The **TPF PROMPT METHOD** is an effective framework for crafting high-quality prompts because it provides a clear, structured approach that simplifies the process of engaging with AI tools. By breaking down the prompt development into six key components, **P**urpose, **R**elevance, **O**utline, **M**odify, **P**ersonalize, and **T**est, this method ensures that non-profits can easily remember and apply each step. The acronym "PROMPT" serves as a memorable guide, allowing users to quickly recall the essential elements needed to create effective prompts. This structured approach not only enhances the quality of AI-generated content but also empowers nonprofits to communicate their unique missions and goals with clarity and impact. Prompts are the key to unlocking the full potential of AI tools, but crafting them effectively requires strategy and precision.

Download the TPF PROMPT METHOD framework in the Nonprofit AI Playbook Toolkit.

	PROCESS	RELEVANCE	OUTLINE	MODIFY	PERSONALIZE	TEST
TASK	Define the purpose of your prompt clearly. What role do you need the AI tool to represent? This step is crucial for guiding the AI in generating relevant content.	Ensure that your prompt is relevant to the context of your nonprofit's mission and goals. Tailor your prompt to reflect the specific needs of your organization.	Provide an outline or structure for the AI to follow. This helps the AI understand the format and organization of the desired output.	Be prepared to modify your prompt based on the AI's initial responses. Iteration is key to refining the output. If the first response isn't what you expected,	Personalize your prompt to make it more engaging and relevant	Test different prompts to see which ones yield the best results. Experimentation is essential for discovering what works best for your needs
GOALS	>What specific information or content do you need? >Are you looking for a summary, a detailed report, or creative content? >Who is the intended audience for the output?	>-Use nonprofit terminology that resonates with your audience. >Reference specific programs, initiatives, or community needs that the AI should consider.	>Key points or topics to cover in the Response. >Specific questions to answer or themes to explore. >Desired length or format (e.g., bullet points, paragraphs, or lists).	>Adjusting the wording of your prompt for clarity. >Adding more context or details to guide the AI. >Asking follow-up questions to delve deeper into the topic.	>Including specific examples or anecdotes related to your nonprofit work. >Using a conversational tone that reflects your organization's voice. >Addressing the AI directly, as if you were having a conversation with a colleague	-Running multiple variations of a prompt to compare outputs. >Analyzing the quality and relevance of the responses generated. >Gathering feedback from team members on the effectiveness of the AI-generated content.

The **TPF PROMPT METHOD** serves as a structured guide to ensure that every AI-generated output, whether it's an **image, brand profile, social media post, blog article, email campaign, grant proposal, or course content**, aligns with your nonprofit's mission and objectives.

By following this method, you will soon consistently develop **clear, relevant, and results-driven prompts** that produce high-quality

AI-generated content. Each step, **Purpose, Relevance, Outline, Modify, Personalize, and Test**, ensures that your prompts are **thorough, strategic, and adaptable**, leading to more impactful results.

Whether you're using AI to generate marketing materials, streamline fundraising efforts, or develop training resources, the **TPF PROMPT METHOD** helps you craft prompts that maximize AI's effectiveness while staying true to your organization's unique voice and goals.

Understanding AI Prompts: How Styles Differ Across Tools

1. **Image Generation Prompt** Template

Use When: Creating AI-generated images for websites, marketing, or social media.

TPF PROMPT Method Applied:

◇ P - Purpose: Clearly define what the image represents.

◇ R - Relevance: Ensure the style fits the intended use (e.g., social media, blog, ad).

◇ O - Outline: Describe key visual elements (colors, lighting, mood, subject).

◇ M - Modify: Adjust for clarity if the initial result isn't as expected.

◇ P - Personalize: Add unique features, like brand colors or themes.

◇ T - Test: Experiment with different styles or details.

Example Image Prompt:

"A diverse group of nonprofit leaders in a modern office, brainstorming ideas with AI-powered tools on a large screen. The setting is well-lit, professional, and futuristic. The color palette includes blues and

greens, representing trust and innovation. Photorealistic, 4K resolution, cinematic lighting."

2. **Brand Profile Prompt** Template

Use When: Crafting nonprofit brand descriptions for websites, brochures, or funding proposals.

TPF PROMPT Method Applied:

- ◇ P - Purpose: Define the goal of the brand profile.
- ◇ R - Relevance: Align messaging with the nonprofit's mission.
- ◇ O - Outline: Provide structure (Mission, Vision, Values, etc.).
- ◇ M - Modify: Refine based on AI output.
- ◇ P - Personalize: Use real success stories or examples.
- ◇ T - Test: Compare multiple versions for effectiveness.

Example Brand Profile Prompt:

"Generate a brand profile for The Philantrepreneur Foundation, a nonprofit that builds nonprofit capacity through education, awareness, and strategic resources. Include a mission statement, vision, core values, and key programs. The tone should be professional yet approachable, and the profile should emphasize impact and innovation in the nonprofit sector."

3. **Social Media Content Prompt** Template

Use When: Generating posts for LinkedIn, Instagram, Facebook, or Twitter/X.

TPF PROMPT Method Applied:

- ◇ P - Purpose: Define the goal (awareness, engagement, donations).
- ◇ R - Relevance: Ensure the content fits the platform's style.

◇ O - Outline: Specify length, tone, hashtags, and CTA.

◇ M - Modify: Tweak for better engagement.

◇ P - Personalize: Add real stories, emojis, or brand elements.

◇ T - Test: Try multiple versions for A/B testing.

Example Social Media Prompt:

"Create an engaging LinkedIn post promoting The Nonprofit AI Playbook. The tone should be inspiring and informative. Start with a compelling hook, highlight how AI can help nonprofits streamline fundraising and operations, and end with a call to action linking to the book. Use hashtags like #NonprofitAI #Fundraising #Innovation."

Example Output:

AI is revolutionizing the nonprofit sector! From automating grant research to predicting donor trends, AI tools are helping mission-driven organizations achieve more with less. Learn how in *The Nonprofit AI Playbook*—your guide to leveraging AI for impact. Grab your copy now!

🖥️ ➡️ NonprofitAIPlaybook.org #NonprofitAI #InnovationForGood #FundraisingTech

4. **Blog & Article Prompts**

Use When: Creating long-form content for nonprofit education, thought leadership, or SEO.

TPF PROMPT Method Applied:

◇ P - Purpose: Define the main topic and target audience.

◇ R - Relevance: Ensure it aligns with your nonprofit's goals and voice.

◇ O - Outline: Provide headings and key points.

◇ M - Modify: Adjust tone, length, and structure if needed.

◇ P - Personalize: Add unique insights, quotes, or statistics.

◇ T - Test: Try variations (e.g., listicle vs. case study).

Example Blog Prompt:

"Write a 1,200-word blog post on how AI can help nonprofits optimize fundraising. Include real-world examples, data-driven insights, and a CTA to learn more at NonprofitAIPlaybook.org. The tone should be informative yet engaging, and the format should include subheadings for easy reading."

5. Email Marketing Prompts (For AI Email Tools like Mailchimp AI, ChatGPT, Copy.ai)

Use When: Writing nonprofit newsletters, fundraising emails, or donor outreach.

TPF PROMPT Method Applied:

◇ P - Purpose: Define the goal (donation, event signup, engagement).

◇ R - Relevance: Align with the nonprofit's messaging and timing.

◇ O - Outline: Include subject line, body text, CTA, and personalization.

◇ M - Modify: Adjust tone, length, or urgency.

◇ P - Personalize: Address recipients by name, reference past engagement.

◇ T - Test: A/B test different versions.

Example Email Prompt:

"Write a donor appreciation email for The Philantrepreneur Foundation. The subject line should be warm and engaging. The email should thank them for their support, highlight the impact of their donation,

and invite them to attend an upcoming webinar on AI for nonprofits. End with a personalized CTA encouraging continued involvement."

Example Subject Line: *You're Making a Difference! A Special Thank You from Us*

6. Grant Writing Prompts (For AI Tools like ChatGPT)

Use When: Drafting grant applications, LOIs, or funding proposals.

TPF PROMPT Method Applied:

◇ P - Purpose: Define the grant's focus and funding goal.

◇ R - Relevance: Align with funder priorities and nonprofit objectives.

◇ O - Outline: Specify sections needed (problem statement, impact, budget).

◇ M - Modify: Refine based on reviewer feedback.

◇ P - Personalize: Include success stories, data, and testimonials.

◇ T - Test: Compare different phrasing for clarity and persuasiveness.

Example Grant Prompt:

"Draft a compelling grant proposal for The Philantrepreneur Foundation's AI for Nonprofits initiative. The proposal should include a strong problem statement, a clear solution, measurable outcomes, and a budget overview. Use persuasive language and emphasize community impact."

7. AI-Assisted Course & Training Content Prompts

Use When: Developing nonprofit training courses, certification programs, or workshop materials.

TPF PROMPT Method Applied:

◇ P - Purpose: Define learning objectives and audience.

◇ R - Relevance: Ensure content is nonprofit-specific.

◇ O - Outline: Provide key modules and topics.

◇ M - Modify: Adjust based on participant feedback.

◇ P - Personalize: Use real-world examples, scenarios, or case studies.

◇ T - Test: Review AI-generated course materials for accuracy and engagement.

Example Course Prompt:

"Create an outline for a 4-module online course titled 'AI for Non-profit Fundraising Success.' Each module should include a lesson summary, key takeaways, and an interactive activity. The tone should be practical and easy to follow."

Here are two prompts we used to create these images.

"Produce a modern black and white vector image that illustrates AI fundamentals with a striking infographic approach. Use bold lines and contrasting shades to depict various tools and their respective tasks. Employ a dynamic layout that encourages flow, with arrows and icons guiding the viewer's eye throughout the composition. Integrate subtle texture to give depth, creating a visually engaging piece suitable for educational purposes."

"Illustrate a vintage-inspired black and white vector image symbolizing AI marketing strategies with a classic megaphone front and center. Utilize bold black and white contrasts to clearly define the different

marketing methods, embedding various tasks around the megaphone in a circular pattern, suggesting a smooth, storytelling motion. Enhance the artwork with an elegant, grainy texture that provides a sense of history, while maintaining a clean, educational clarity that invites viewers to explore the intricacies of marketing techniques."

Organizational Prompts Examples

EXAMPLE 1: ORGANIZATION HEALING HEROES SERVING VETERANS WITH PTSD

	PROCESS	TASK	GOAL
P	PURPOSE	You are an expert at creating Brand Profiles. We need to develop a Brand Profile for Healing Heroes that will serve as the foundation for all future communications.?	Use our mission of 'empowering veterans with PTSD to reclaim their lives through compassionate support and community-driven programs' to create a clear and compelling Brand Profile.
R	RELEVANCE	Ensure the Brand Profile is relevant to our mission and resonates with our target audience of veterans, their families, and supporters.	Tailor the content to reflect the specific needs of veterans with PTSD, emphasizing programs like peer support groups, therapy sessions, and family counseling.
O	OUTLINE	Provide a structure for the Brand Profile that includes the following sections: • **Mission Statement**: 'Healing Heroes empowers veterans with PTSD to reclaim their lives through compassionate support and community-driven programs.' • **Core Values**: Compassion, Empowerment, Community, Respect. • **Target Audience**: Veterans with PTSD, their families, and supporters of veteran advocacy.	Ensure the profile is comprehensive and easy to follow, serving as a reference for all communications."

		• **Key Programs**: Peer support groups, therapy sessions, family counseling, and advocacy initiatives. Brand Colors: Navy blue (trust and loyalty), olive green (peace and healing). Fonts: Arial for body text, Georgia for headings. Imagery Preferences: Images of veterans engaging in therapeutic activities, community events, and support groups.	
M	MODIFY	Review the initial draft of the Brand Profile and make adjustments to ensure it aligns with our tone and messaging.	Refine the content to include specific examples, such as success stories from veterans who have benefited from our programs, and ensure the tone is conversational and approachable."
P	PERSON-ALIZE	Add specific examples or anecdotes to make the Brand Profile engaging and relatable.	Include a story about John, a veteran who struggled with PTSD for years until he found Healing Heroes. Through our therapy sessions and community programs, John has rebuilt his life and now volunteers to help others. His story embodies our mission and inspires hope."
T	TEST	Test the Brand Profile with stakeholders, including veterans, caregivers, and donors, to gather feedback.	Ensure the profile resonates with the audience and accurately represents our organization. Use their input to refine the content and make it more impactful."

EXAMPLE 2: PAWSITIVE FUTURES MARKETING CAMPAIGN FOR DOG DAYS IN THE PARK FUNDRAISER

	PROCESS	TASK	GOAL
P	PURPOSE	You are an expert at creating marketing campaigns. We need to develop a campaign for our annual 'Dog Days In The Park' fundraiser.	Promote the event to raise funds for dog rescue efforts, emphasizing the fun activities and the impact of every ticket sold."
R	RELE-VANCE	Ensure the campaign messaging aligns with our mission of rescuing and re-homing dogs in need.	Tailor the content to resonate with dog lovers, families, and potential adopters,

			highlighting how the event supports our rescue efforts."
O	OUTLINE	Provide a structure for the campaign that includes: 1. **Social Media Posts**: Fun, engaging posts with photos of adoptable dogs and event details. 2. **Email Campaign**: A series of emails with event highlights, ticket links, and success stories. 3. **Flyers and Posters**: Eye-catching visuals for local distribution. 4. **Event Page**: A dedicated webpage with event details, ticket sales, and a donation link	Ensure the campaign is cohesive and easy to follow across all platforms."
M	MODIFY	Review the initial campaign materials and make adjustments based on feedback or early results.	Optimize the campaign to maximize engagement and participation. For example, if ticket sales are low, add a limited-time discount for early bird tickets and share more success stories to inspire participation."
P	PERSON-ALIZE	Include specific details about the event and its impact to make the campaign feel personal and exciting	Feature Bella, a rescue dog who found her forever home after last year's event. Her story will show how the fundraiser directly impacts dogs' lives, encouraging attendees to support the cause."
T	TEST	Test different versions of the campaign materials to identify the most effective content	Create two versions of the email campaign: one focusing on event activities and another highlighting the impact of donations. Compare open rates and ticket sales to determine which approach resonates most with our audience."

PRO TIP: In both examples, the OUTLINE section is the most detailed and requires the most attention, think of it as the blueprint for your prompt. The MODIFY section is where the magic happens, allowing you to refine and perfect the results. Don't be afraid to ask the AI, *"Do you have any questions for me?"* You'll be amazed at how intuitive AI can be, it often asks questions you hadn't considered, helping you uncover new insights and improve your prompts even further!

Question: What is one task or project in your organization where you could apply the TPF PROMPT Method?

Note Space: Draft a sample prompt for that task or project.

_____

_____

_____

_____

_____

_____

_____

_____

_____

_____

_____

_____

_____

_____

_____

_____

_____

_____

_____

_____

_____

CHAPTER 6

AI for Marketing and Communications

A strong marketing plan is essential for establishing a nonprofit's brand, mission, and outreach strategy, and it should ideally be developed alongside or even before a fundraising plan. This approach ensures that the organization has a solid foundation for attracting support and resources. This chapter will outline essential elements to equip nonprofits with the knowledge and tools to leverage artificial intelligence (AI) for effective marketing and engagement.

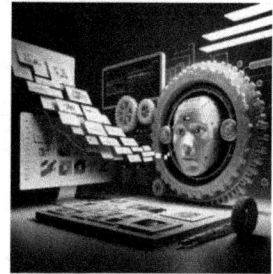

The Importance of Marketing Before Fundraising

A well-crafted marketing plan serves as the backbone of a nonprofit's outreach efforts. It defines the organization's brand identity, articulates its mission, and outlines strategies for engaging with the community and potential donors. A strong marketing strategy not only raises awareness about the nonprofit's cause but also builds trust and credibility among stakeholders.

For instance, consider a nonprofit focused on environmental conservation. By developing a marketing plan that highlights its mission, showcases success stories, and engages the community through educational campaigns, the organization can create a strong emotional connection with potential donors. This connection is crucial, as donors are more likely to contribute to causes they feel personally invested in.

Moreover, a marketing plan can help identify target audiences, allowing nonprofits to tailor their messaging and outreach efforts effectively. By understanding who their supporters are, nonprofits can create campaigns that resonate with their values and interests, ultimately leading to increased engagement and donations.

Ethical Considerations in AI

As nonprofits increasingly adopt AI tools for marketing, it is essential to address the ethical considerations that accompany their use.

- **Bias:** AI algorithms can inadvertently perpetuate biases present in the data they are trained on. Nonprofits must be vigilant in selecting diverse data sets to ensure fair representation and avoid reinforcing stereotypes. For example, if a nonprofit uses AI to analyze donor demographics, it should ensure that the data reflects a broad spectrum of the community it serves.
- **Transparency:** Transparency in AI processes is vital for building trust with stakeholders. Nonprofits should communicate how AI tools are used in their marketing efforts, including how

data is collected and analyzed. This openness fosters accountability and helps stakeholders understand the rationale behind marketing decisions.

- **Data Privacy:** Protecting the privacy of donors and stakeholders is paramount. Nonprofits must adhere to data protection regulations, such as the General Data Protection Regulation (GDPR) and the California Consumer Privacy Act (CCPA). Implementing robust data security measures and clearly communicating privacy policies can help build trust and encourage donor engagement.

AI Tools for Nonprofit Marketing

Nonprofits can harness various AI tools to enhance their marketing efforts, streamline processes, and improve engagement.

- **AI Playbook Toolkit:** This toolkit empowers nonprofits to create compelling content, manage social media campaigns, and analyze engagement metrics effortlessly. For example, the toolkit may include templates for social media posts, email newsletters, and blog articles, allowing nonprofits to maintain a consistent brand voice while saving time on content creation. Additionally, built-in analytics features can help organizations track engagement metrics, enabling them to refine their strategies based on real-time data.
- **Galaxy.ai:** This advanced analytics tool provides nonprofits with insights to optimize their marketing strategies. By analyzing data from various sources, Galaxy.ai can identify trends, audience preferences, and campaign performance. For instance, a nonprofit may discover that its social media posts featuring

personal stories receive higher engagement than generic up-dates. Armed with this knowledge, the organization can adjust its content strategy to focus on storytelling, ultimately enhancing its outreach efforts.

- **QRCodeChimp:** Customizable QR codes can drive traffic to campaigns and enhance engagement. Nonprofits can use QR codes on printed materials, such as flyers or brochures, to direct potential supporters to their websites or donation pages. For example, a nonprofit hosting an event could include a QR code on its promotional materials, allowing attendees to easily access event details or make donations on the spot.

- **AI Workstation (Dr Boyd bot)**

Social Media Management and Analytics with AI

AI plays a significant role in automating social media management, allowing nonprofits to focus on creating meaningful content. AI tools can schedule posts, analyze audience engagement, and suggest optimal posting times based on historical data.

For instance, a nonprofit may use an AI-driven social media management tool to analyze which types of posts generate the most engagement. By identifying patterns in audience behavior, the

organization can tailor its content strategy to maximize reach and impact. Additionally, AI can help nonprofits monitor social media conversations, enabling them to respond promptly to inquiries or comments, further enhancing their engagement with supporters.

AI-Driven Email Marketing and A/B Testing

Email marketing remains a powerful tool for nonprofits, and AI can significantly enhance its effectiveness. AI-driven email marketing platforms can analyze donor behavior and preferences, allowing organizations to segment their audiences and personalize content. For example, a nonprofit may use AI to identify donors who have previously contributed to specific campaigns and tailor email messages to encourage further support for similar initiatives.

A/B testing is another valuable strategy that AI can facilitate. By testing different subject lines, content formats, or calls to action, nonprofits can determine which elements resonate most with their audience. For instance, a nonprofit might find that emails with personalized subject lines yield higher open rates than generic ones. Armed with this knowledge, the organization can refine its email marketing strategy to improve engagement and conversion rates.

Measuring ROI of AI-Powered Campaigns

Measuring the return on investment (ROI) of AI-powered marketing campaigns is essential for understanding their effectiveness. Nonprofits should establish key performance indicators (KPIs) to evaluate the success of their efforts. Common KPIs include engagement rates, conversion rates, and overall fundraising results.

To calculate ROI, nonprofits can compare the costs associated with their marketing campaigns to the revenue generated from donations or other sources. For example, if a nonprofit invests $1,000 in an AI-driven social media campaign that results in $5,000 in donations, the ROI would be 400%. This data-driven approach allows organizations to assess the impact of their marketing strategies and make informed decisions about future investments.

Leveraging the Google Ad Grant

One of the most powerful tools available to nonprofits is the Google Ad Grant, which provides up to $10,000 per month in free advertising. This grant can significantly amplify your organization's reach, helping you attract donors, volunteers, and supporters. For a detailed guide on how to apply, set up, and optimize your Google Ad Grant campaigns, see Chapter 6: Maximizing Impact with the Google Ad Grant.

Question: How will you ensure that your nonprofit's use of AI in marketing respects donor privacy and avoids bias?

Note Space: Write down 1-2 steps you'll take to address ethical concerns in your AI-driven marketing strategies.

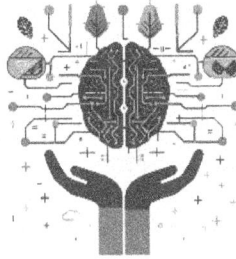

Chapter 7

Maximizing Impact with the Google Ad Grant

Google Ad Grants The **Google Ad Grant** is one of the most powerful tools available to nonprofits, offering up to **$10,000 per month in free advertising** on Google's search platform. This grant allows organizations to reach new audiences, drive traffic to their websites, and amplify their impact, all without spending a dime on advertising. There are some fundamental things you need to know and learn before running off to get a Google Ad Grant and we will provide as much information as possible right here. What we will cover is:

1. **What the Google Ad Grant is**, its history, and how it works.

2. **How to qualify for the grant** and ensure your organization meets all eligibility requirements.

3. **Step-by-step instructions** for applying and setting up your first campaign.

4. **How to maintain compliance** with Google's rules to keep your grant active.

5. **How to use AI tools** to streamline campaign management, optimize performance, and save time.

6. **Real-world success stories** of nonprofits that have leveraged the Google Ad Grant to achieve their goals.

By the end of this chapter, you'll have the knowledge and tools to unlock the full potential of the Google Ad Grant and take your nonprofit's marketing efforts to the next level.

What is the Google Ad Grant?

The Google Ad Grant program was launched in 2003 and has helped thousands of nonprofits worldwide increase their visibility and achieve their missions. To date, Google has awarded over **$10 billion in advertising credits** to eligible organizations, making it one of the largest philanthropic initiatives in the digital space.

Those awarded a Google Ad Grant are eligible for **up to $10,000 per month in free advertising credits** on Google's search platform. These credits can be used to create text-based ads that appear in Google search results, helping organizations reach new audiences, drive traffic to their websites, and amplify their impact.

How It Works

- **Credits**: Nonprofits receive a monthly budget of $10,000 in advertising credits, which can be used to create and run text ads on Google Search.

- **Ad Types**: Google Ad Grant campaigns are limited to **text ads**, which appear in search results and include a headline, description, and display URL.
- **Eligible Campaigns**: Ads must promote the nonprofit's mission, programs, or services. Prohibited content includes gambling, alcohol, and political advocacy.

Pro Tip: While the Google Ad Grant is a powerful tool, it's important to understand its limitations and requirements to make the most of it.

Eligibility Requirements

Before diving into the application process, it's important to ensure your organization meets Google's eligibility criteria. The Google Ad Grant is available to **501(c)(3) nonprofits** in eligible countries, but there are a few additional requirements to keep in mind:

1. **Valid Nonprofit Status**: Your organization must hold a valid 501(c)(3) status (or equivalent in your country) and be registered with TechSoup or a local partner.
2. **High-Quality Website**: Your website must be secure (HTTPS), have clear navigation, and provide valuable content about your mission and programs.
3. **Compliance with Google's Policies**: Your organization must adhere to Google's nonprofit policies, which prohibit certain types of content and advertising practices.

Pro Tip: Before applying, review your website and ensure it meets Google's standards. If you're unsure, our **10K Money Guide** includes a checklist to help you prepare.

ALL ROADS
LEAD TO YOUR

How to Apply

Applying for the Google Ad Grant is a multi-step process, but with the right preparation, it can be straightforward and stress-free. Here's an overview of the steps involved:

1. **Register with TechSoup**: If you haven't already, sign up for a TechSoup account and validate your nonprofit status.

2. **Create a Google for Nonprofits Account**: This account will serve as your hub for accessing the Google Ad Grant and other Google nonprofit tools.

3. **Complete the Google Ad Grant Application**: Fill out the application form, providing details about your organization, mission, and website.

4. **Set Up Google Ads**: Once approved, you'll need to create your first campaign in Google Ads. This includes selecting keywords, writing ad copy, and setting a budget (up to $329 per day).

Maintaining Compliance

Once you've been approved for the Google Ad Grant, maintaining compliance is essential to keep your grant active. Google has specific rules and requirements that nonprofits must follow, and failure to comply can result in suspension or loss of the grant. Here are the key compliance rules to keep in mind:

1. **Active Account Management**: Your Google Ads account must be actively managed, with regular updates to campaigns, keywords, and ad copy.

2. **Minimum Performance Standards**: Your account must maintain a **5% click-through rate (CTR)** and use high-quality, relevant keywords.

3. **Prohibited Content**: Avoid promoting prohibited content, such as gambling, alcohol, or political advocacy.

4. **Monthly Reporting**: Submit monthly reports to Google to demonstrate your account's activity and performance.

> **Pro Tip**: Set up a compliance checklist and schedule regular reviews of your account to ensure you're meeting Google's requirements.

Implementing Ad Campaigns - Your AI-Powered Ad Grant Manager

Entry-Level Ad Managers: Typically charge $500–$800 per month for basic campaign setup and management.

Mid-Level Ad Managers: Charge $800–$1,200 per month for more advanced services, including optimization and reporting.

Expert Ad Managers: Can charge $1,200–$1,500+ per month for comprehensive campaign management, including strategy development and A/B testing.

For nonprofits, these costs can be prohibitive, making AI-powered solutions an attractive alternative.

Managing a Google Ad Grant campaign can be a daunting task, especially for nonprofits with limited budgets and resources. Hiring a professional ad manager can cost anywhere from **$500 to $1,500 per month**, a price tag that's often out of reach for smaller organizations. But what if you could have a **dedicated Ad Grant Manager**—one that's available 24/7, never misses a detail, and doesn't charge a dime? Enter **AI**.

By leveraging AI tools, nonprofits can streamline every aspect of their Google Ad Grant campaigns—from setting up and optimizing ads to monitoring performance and ensuring compliance. AI acts as your **"Ad Grant Manager,"** providing expert guidance and actionable insights at a fraction of the cost.

Google's Performance Max AI-Powered Solution - a Game-Changer for Nonprofits

In the ever-evolving world of digital advertising, Google has introduced *Performance Max*, a cutting-edge, AI-powered campaign manager designed to maximize results across all of Google's platforms. Think of Performance Max as a self-driving car for your Google Ads: you set the destination (your goal), and the AI takes care of the rest, choosing the best route, adjusting speed, and avoiding obstacles. Even though Google Ad Grant campaigns are limited to text ads, Performance Max's AI-driven optimization ensures your ads are shown to the right audience at the right time, maximizing your impact. By automating the complex process of ad management, Performance Max allows nonprofits to focus on their mission while still leveraging the power of Google Ads.

For nonprofits leveraging the Google Ad Grant, Performance Max offers a powerful way to reach more donors, volunteers, and supporters while optimizing your budget and time.

However, it's important to note that Google Ad Grant campaigns are limited to text ads, even though Performance Max can accommodate other ad types like images and videos. This means nonprofits must focus on crafting compelling text-based ads while still benefiting from the AI-driven optimization that Performance Max provides.

For complete instructions on how to set up a Performance Max account, check out the Nonprofit AI Playbook Toolkit at **https://NonprofitAIPlaybook.org**

What is Performance Max?

Performance Max is a goal-based campaign type that uses Google's AI to optimize your ads across all of its channels, including Search, Display, YouTube, Gmail, and Maps. Unlike traditional campaigns, which require manual setup and management for each channel, Performance Max automates the process, allowing you to focus on your goals while Google's AI handles the rest.

Key features of Performance Max include:

- Automated Optimization: AI analyzes your campaign data in real-time to adjust bids, placements, and ad creatives for maximum performance.
- Cross-Channel Reach: Your ads are shown across Google's entire network, ensuring you reach your audience wherever they are.
- Goal-Based Campaigns: You set the goal (e.g., donations, event sign-ups, website traffic), and AI works to achieve it.
- Dynamic Creative Optimization: AI tests different combinations of headlines, descriptions, and images to find the most effective ad variations.

How Nonprofits Can Use Performance Max

While Performance Max supports multiple ad types, Google Ad Grant campaigns are limited to text ads. Here's how nonprofits can use Performance Max to amplify their impact within these constraints:

1. **Driving Donations**
   - Goal: Increase online donations.
   - Setup: Upload your donation page URL, select "Conversions" as the goal, and provide text-based assets like headlines and descriptions.

- Outcome: AI optimizes your text ads to reach potential donors and drive them to your donation page, maximizing contributions.

2. **Promoting Events**
   - Goal: Boost event registrations.
   - Setup: Upload your event page URL, select "Conversions" as the goal, and provide text-based assets like event details and a call-to-action (e.g., "Register Now").
   - Outcome: AI promotes your event across Google's platforms, ensuring it reaches the right audience and drives registrations.

3. **Recruiting Volunteers**
   - Goal: Increase volunteer sign-ups.
   - Setup: Upload your volunteer sign-up page URL, select "Conversions" as the goal, and provide text-based assets like volunteer opportunities and testimonials.
   - Outcome: AI targets individuals interested in volunteering and directs them to your sign-up page, growing your volunteer base.

4. **Raising Awareness**
   - Goal: Increase website traffic and awareness.
   - Setup: Upload your homepage URL, select "Traffic" as the goal, and provide text-based assets like mission statements and key programs.
   - Outcome: AI drives traffic to your website, helping you raise awareness about your mission and programs.

Optimizing Your Campaign with AI

Once your campaign is live, AI can help you monitor performance, make data-driven adjustments, and ensure you're getting the most out of your Google Ad Grant. Here's how AI can act as your **"Ad Grant Manager"** during the optimization phase:

1. Performance Monitoring

- **AI Application**: AI tools like Google Analytics or Tableau can track key metrics such as click-through rate (CTR), conversion rate, and cost per click (CPC).
- **Example Prompt**: *"Analyze the performance of our Google Ads campaign and identify areas for improvement."*
- **Outcome**: AI provides actionable insights, such as underperforming keywords or ads, helping you refine your strategy.

2. A/B Testing

- **AI Application**: AI can help you create and test multiple versions of ad copy, headlines, and calls-to-action to determine what resonates best with your audience.
- **Example Prompt**: *"Generate three variations of our ad copy for A/B testing, focusing on different calls-to-action."*
- **Outcome**: AI identifies the most effective version, improving campaign performance.

3. Budget Optimization

- **AI Application**: AI can analyze your campaign spending and recommend adjustments to maximize your budget.
- **Example Prompt**: *"Review our Google Ads budget and suggest ways to allocate funds more effectively."*
- **Outcome**: AI identifies areas where you can reduce costs or reallocate funds to higher-performing campaigns.

4. Compliance Monitoring

- **AI Application**: AI can help you stay compliant with Google's Ad Grant requirements by flagging potential issues, such as low CTR or prohibited keywords.
- **Example Prompt**: *"Check our Google Ads account for compliance issues and suggest corrective actions."*
- **Outcome**: AI ensures your account meets Google's standards, reducing the risk of suspension.

5. Reporting and Insights

- **AI Application**: AI can generate detailed reports on campaign performance, highlighting key metrics and trends.
- **Example Prompt**: *"Create a monthly performance report for our Google Ads campaign, including CTR, conversions, and ROI."*
- **Outcome**: AI provides clear, data-driven insights to inform your decision-making and demonstrate impact to stakeholders.

Success Stories

1. **Healing Heroes: Boosting Donor Engagement**
   Healing Heroes, a nonprofit supporting veterans with PTSD, used the Google Ad Grant to promote their peer support programs. By leveraging AI to craft compelling ad copy and target the right audience, they saw a **40% increase in website traffic** and a **25% boost in donations** within three months.

2. **Pawsitive Futures: Finding Forever Homes**
   Pawsitive Futures, a dog rescue organization, used the Google Ad Grant to promote their adoption events and fundraising campaigns. With AI's help, they optimized their keywords and

ad placements, resulting in a **30% increase in event attendance** and a **50% rise in adoption inquiries**.

3. **Green Earth Initiative: Raising Awareness**

   Green Earth Initiative, an environmental nonprofit, used the Google Ad Grant to raise awareness about their conservation programs. AI-powered tools helped them create engaging ad content and track campaign performance, leading to a **60% increase in volunteer sign-ups** and a **20% boost in social media followers**.

Don't Miss Out

The Google Ad Grant is a game-changer for nonprofits, offering up to **$10,000 per month in free advertising** to amplify your mission and reach new audiences. Yet, many organizations miss out on this opportunity, leaving **$10,000 on the table every month**, money that could be used to drive donations, recruit volunteers, and grow your community.

By leveraging AI as your **"Ad Grant Manager,"** you can unlock the full potential of this grant, saving thousands on ad management costs while maximizing your impact. Whether you're promoting a fundraising event, raising awareness about your mission, or recruiting volunteers, the Google Ad Grant—combined with AI's power—can help you achieve your goals.

Don't let this opportunity pass you by. Start your Google Ad Grant journey today and see how AI can transform your nonprofit's marketing efforts.

Question: What specific goal (e.g., donations, event sign-ups, volunteer recruitment) would you focus on for your first Google Ad Grant campaign?

Note Space: Outline the steps you'll take to set up and optimize your campaign.

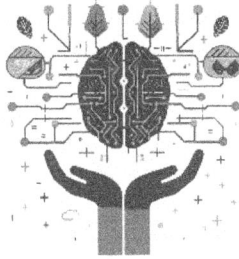

Chapter 8

AI for Fundraising

It's a fact of life; most nonprofits are concerned about fundraising. Therefore, we aim to demonstrate how artificial intelligence (AI) can significantly enhance donor engagement and improve fundraising success for nonprofits. In an increasingly competitive landscape, leveraging AI tools and strategies can help organizations identify potential donors, personalize communication, and predict fundraising outcomes, ultimately leading to more effective campaigns and increased revenue.

Solving Common Fundraising Mistakes with AI

Let's address the elephant in the room first – the mistakes made by so many nonprofits as they implement fundraising their strategies. We'll just list some common mistakes and add AI solutions.

- Mistake: Lack of Donor Engagement

- AI Solution: Use AI-powered chatbots to engage donors in real-time, answer questions, and provide personalized updates.
- Actionable Tip: Implement a chatbot on your website to guide donors through the giving process.
- Mistake: Inefficient Donor Segmentation
  - AI Solution: Use AI to segment donors based on giving history, interests, and engagement levels.
  - Actionable Tip: Create targeted campaigns for each segment (e.g., first-time donors, recurring donors, major gift prospects).
- Mistake: Poor Timing of Fundraising Appeals
  - AI Solution: Use predictive analytics to identify the best times to reach out to donors.
  - Actionable Tip: Schedule fundraising emails and social media posts based on AI-generated insights.
- Mistake: Generic Messaging
  - AI Solution: Use AI to generate personalized messages that resonate with individual donors.
  - Actionable Tip: Craft tailored emails and social media posts using AI tools like ChatGPT or Jasper.

Ethical Considerations in AI Fundraising

As nonprofits explore the potential of AI to enhance their fundraising efforts, it's natural to have concerns about the ethical implications of using these technologies. Some may worry that AI could feel impersonal, invasive, or even unethical. However, when used responsibly, AI can be a powerful tool for building stronger donor relationships and amplifying your impact.

This section addresses common ethical concerns, **data privacy, transparency, and bias**, and provides practical steps to ensure your use of AI aligns with your organization's values and mission.

Data Privacy: Protecting Donor Information

The Concern: Nonprofits often worry that using AI could compromise donor privacy or lead to misuse of sensitive data.

Why It Matters: Donors trust you with their personal information, and maintaining that trust is critical to your organization's success.

How to Address It:

- **Collect Only What You Need**: Limit the data you collect to what's necessary for your fundraising efforts. For example, avoid asking for sensitive information unless it's essential.
- **Use Secure Tools**: Ensure the AI tools you use comply with data protection regulations like **GDPR** (General Data Protection Regulation) or **CCPA** (California Consumer Privacy Act).
- **Encrypt Data**: Use encryption to protect donor data both in transit and at rest.
- **Be Transparent**: Clearly communicate how donor data will be used and stored, and provide an easy way for donors to opt out of data collection.

Pro Tip: Regularly audit your data practices to ensure compliance with privacy laws and best practices.

Transparency: Building Trust with Donors

The Concern: Donors may feel uneasy if they don't understand how AI is being used in your fundraising efforts.

Why It Matters: Transparency fosters trust, and trust is the foundation of strong donor relationships.

How to Address It:

- **Explain How AI Works**: Use simple, non-technical language to explain how AI helps your organization. For example, "We use AI to personalize our communications and ensure you receive updates that matter most to you."

- **Disclose AI Use**: Be upfront about when and how AI is used in your fundraising campaigns. For instance, if you're using AI to analyze donor behavior, let donors know.

- **Provide Control**: Give donors the option to opt out of AI-driven processes, such as personalized messaging or automated outreach.

- **Share Success Stories**: Highlight how AI has helped your organization achieve its goals, such as increasing donations or improving program efficiency.

Pro Tip: Create a dedicated page on your website that explains your AI practices and addresses common questions.

Bias: Ensuring Fairness and Inclusion

The Concern: AI systems can inadvertently perpetuate biases, leading to unfair or exclusionary practices.

Why It Matters: Nonprofits are committed to equity and inclusion, and biased AI could undermine those values.

How to Address It:

- **Use Diverse Data**: Ensure the data used to train AI models is representative of your donor base and community.
- **Audit for Bias**: Regularly review AI outputs to identify and address any biases. For example, check if certain donor segments are being overlooked or unfairly targeted.
- **Involve Stakeholders**: Engage staff, donors, and community members in the development and implementation of AI tools to ensure they align with your organization's values.
- **Choose Ethical Tools**: Select AI tools that prioritize fairness and inclusion and avoid those with a history of biased outcomes.

Pro Tip: Partner with AI experts or consultants who specialize in ethical AI to ensure your tools and practices are fair and inclusive.

Reducing Fears About AI

Many nonprofits worry that using AI might feel impersonal or unethical. Here's how to address those fears:

- **AI Enhances, Not Replaces**: Emphasize that AI is a tool to enhance human efforts, not replace them. For example, AI can automate repetitive tasks, freeing up your team to focus on building personal relationships with donors.
- **Donor-Centric Approach**: Use AI to create more personalized, meaningful experiences for donors, such as tailored messages or targeted outreach.
- **Ethical AI Practices**: By following ethical guidelines, like protecting data privacy, being transparent, and addressing bias, you can ensure your use of AI aligns with your mission and values.

Best Practices for Ethical AI Fundraising

1. **Develop an AI Policy**: Create a clear policy that outlines how your organization will use AI, including guidelines for data privacy, transparency, and bias.

2. **Train Your Team**: Educate staff and volunteers on ethical AI practices and how to use AI tools responsibly.

3. **Engage Donors**: Involve donors in conversations about AI, addressing their concerns and explaining how it benefits your mission.

4. **Monitor and Improve**: Continuously evaluate your AI practices and make improvements as needed to ensure they remain ethical and effective.

Using AI in fundraising doesn't have to be unethical or impersonal. By prioritizing **data privacy**, **transparency**, and **fairness**, nonprofits can harness the power of AI to build stronger donor relationships and achieve their mission.

Remember: AI is a tool, and like any tool, its impact depends on how it's used. By adopting ethical practices, you can ensure that AI enhances your fundraising efforts while staying true to your organization's values.

AI for Donor Identification and Segmentation

One of the most critical aspects of successful fundraising is identifying and understanding your donor base. AI can streamline this process by analyzing vast amounts of data to identify potential donors and segment them based on various criteria.

- **Data Analysis for Donor Identification:**
  AI tools can analyze historical donation data, social media activity, and demographic information to identify individuals who are likely to support your cause. For example, a nonprofit focused on education might use an **AI Search Engine** to gather insights from various online sources about potential donors, identifying patterns such as age, income level, and geographic location. This analysis can reveal potential new donors who share similar characteristics.

- **Segmentation for Targeted Campaigns:**
  Once potential donors are identified, AI can help segment them into specific groups based on their interests, giving history, and engagement levels. Tools like **Custom GPTs** can assist in generating tailored messaging for different donor segments. For instance, a health-focused nonprofit might segment its donors into categories such as "first-time donors," "recurring donors," and "major gift prospects." This segmentation allows organizations to tailor their fundraising strategies to each group, ensuring that messaging resonates with the unique motivations of each donor segment.

SUPERDONOR FRAMEWORK

Have You Heard of the SuperDonor Framework?

For years, nonprofits have relied on the Donor Engagement Continuum—a model that measures donor "value" based on monetary contributions and giving frequency. While this approach has its merits, it falls short in one critical area: true engagement. The continuum focuses on how much donors give and how often, but it doesn't address the deeper, more meaningful ways donors can connect with and support your mission.

Enter the SuperDonor Framework (SDF), a revolutionary approach inspired by Pat Flynn's SuperFan concept. The SDF shifts the focus from monetary value to people-first engagement, recognizing that donors are more than just financial contributors—they are advocates, volunteers, and ambassadors for your cause.

The Engagement Gap

The traditional Donor Engagement Continuum identifies five levels of giving:

1. First-Time Donors: Those who make an initial contribution.
2. Repeat Donors: Those who give multiple times.
3. Sustaining Donors: Those who commit to recurring donations.
4. Major Donors: Those who contribute significant sums.
5. Legacy Donors: Those who include your organization in their estate plans.

While this model provides a useful framework for tracking financial contributions, it overlooks the multifaceted ways donors can engage with your organization. True engagement goes beyond writing a check—it includes volunteering, advocating, sharing skills, participating in events, and even providing constructive feedback.

The SuperDonor Framework: Redefining Engagement

The SuperDonor Framework flips the script by prioritizing engagement over monetary value. It's about creating a donor journey that fosters long-term relationships and transforms first-time donors into passionate, committed advocates, your SuperDonors.

SuperDonors are more than just donors; they are individuals who:

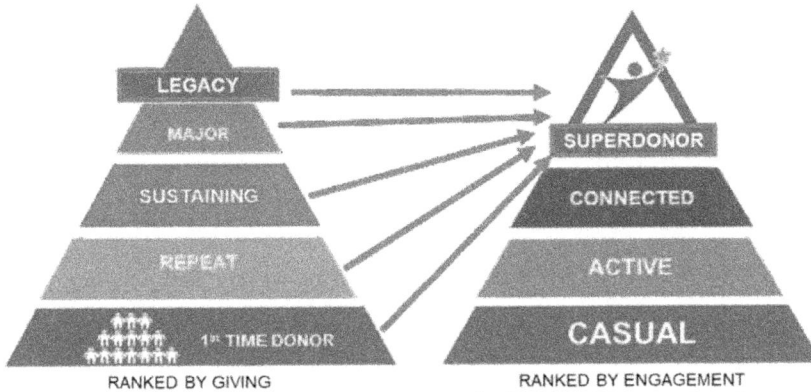

RANKED BY GIVING — RANKED BY ENGAGEMENT

- Believe deeply in your cause and align with your mission.
- Give consistently and generously, whether through financial contributions, time, or skills.
- Advocate passionately for your organization, spreading awareness and inspiring others to get involved.
- Stay committed for the long haul, becoming pillars of support and stability for your mission.

The SDF is built on the idea that engagement is the key to donor retention and growth. By focusing on building relationships and creating meaningful connections, nonprofits can cultivate a community of SuperDonors who are fully invested in their success.

Where AI Comes In

While the **SuperDonor Framework** provides the roadmap for building engagement, Artificial Intelligence (AI) offers the tools to make it happen efficiently and effectively. By aligning the **Super-Donor Framework** funnel with AI strategies, nonprofits can create a seamless, donor-centric process that transforms unaware prospects into committed SuperDonors. From building awareness to fostering advocacy, AI tools provide the insights and automation needed to make every step of the journey more effective and efficient.

Here's a **step-by-step process** that aligns with the **SuperDonor Framework (SDF) funnel** from **Unaware, Aware, Engaged, Connected, and Committed (SuperDonor)**, and integrates AI strategies at each stage. This process is designed to guide nonprofits transforming prospects into committed SuperDonors while leveraging AI tools to enhance efficiency and effectiveness.

Step 1: Unaware → Aware (Building Awareness)

Goal: Introduce your organization and mission to potential supporters.

1. **Identify Your Audience**: Use **AI-driven data analysis tools** to analyze demographic and behavioral data and identify potential supporters who align with your mission.
2. **Craft Compelling Messages**: Use **AI content generation tools** to create engaging content that highlights your mission and impact.
3. **Leverage Multiple Channels**: Use **AI-powered social media management tools** to schedule and optimize posts, ensuring your message reaches the right audience.

4. **Track Engagement**: Use **AI analytics tools** to monitor website traffic and social media engagement, identifying which messages resonate most.

AI Tools: Data analysis tools, content generation tools, social media management tools, analytics tools

Step 2: Aware → Engaged (Encouraging Initial Involvement)

Goal: Encourage prospects to take their first step toward involvement, such as signing up for a newsletter or attending an event.

1. **Personalize Outreach**: Use **AI email marketing tools** to send personalized emails based on the recipient's interests and behavior.
2. **Host Virtual Events**: Use **AI event management tools** to host webinars or workshops that educate prospects about your mission.
3. **Use Chatbots**: Implement **AI-powered chatbots** to answer questions and guide prospects toward involvement.
4. **Collect Data**: Use **AI engagement tracking tools** to track metrics like email open rates and event attendance, refining your strategies.

AI Tools: Email marketing tools, event management tools, chatbots, engagement tracking tools

Step 3: Engaged → Connected (Building Relationships)

Goal: Deepen the connection by encouraging prospects to become donors, volunteers, or advocates.

1. **Segment Your Audience**: Use **AI donor segmentation tools** to group prospects based on their level of engagement and interests.

2. **Personalize Communication**: Use **AI content generation tools** to craft tailored messages that speak to each segment's motivations and values.

3. **Simplify the Donation Process**: Use **AI website optimization tools** to ensure your donation page is user-friendly and mobile-responsive.

4. **Encourage Volunteering**: Use **AI volunteer matching tools** to match volunteers with opportunities that align with their skills and interests.

AI Tools: Donor segmentation tools, content generation tools, website optimization tools, volunteer matching tools

Step 4: Connected → Committed (Fostering Commitment)

Goal: Transform connected supporters into committed SuperDonors who are deeply invested in your mission.

1. **Identify High-Potential Donors**: Use **AI predictive analytics tools** to identify supporters most likely to become major or recurring donors.

2. **Personalize Cultivation Strategies**: Use **AI content generation tools** to create tailored outreach plans for each high-potential donor.

3. **Involve Donors in Decision-Making**: Use **AI meeting facilitation tools** to host virtual meetings with donors, inviting them to join your board or committees.

4. **Showcase Impact**: Use **AI design tools** to create personalized impact reports that demonstrate how their contributions are making a difference.

AI Tools: Predictive analytics tools, content generation tools, meeting facilitation tools, design tools

Step 5: Committed → SuperDonor (Cultivating Ambassadors)

Goal: Turn committed donors into passionate advocates and ambassadors for your cause.

1. **Encourage Advocacy**: Use **AI advocacy tools** to identify donors who are most likely to advocate for your cause and provide them with resources to share your mission.

2. **Facilitate Peer-to-Peer Fundraising**: Use **AI fundraising tools** to help SuperDonors create personal fundraising campaigns.

3. **Recognize and Celebrate**: Use **AI recognition tools** to track and celebrate SuperDonors' contributions, whether through personalized thank-you messages or public recognition.

4. **Foster Long-Term Relationships**: Use **AI communication tools** to maintain regular communication with SuperDonors, keeping them informed and engaged with your mission.

AI Tools: Advocacy tools, fundraising tools, recognition tools, communication tools

Ready to get started? Download our **AI Fundraising Toolkit** or enroll in our **SuperDonor Framework Training** to learn how to implement these strategies in your organization.

Personalized Donor Communication Using AI

The SuperDonor Framework is built on the foundation of communication as the key to building a relationship. The key to effective donor communication is *personalization*. AI can enhance this aspect by enabling nonprofits to create tailored messages that resonate with individual donors, ultimately fostering stronger relationships and increasing the likelihood of contributions.

- **Dynamic Content Creation:**
  AI-driven tools like **AI Content Detector** can analyze donor data to create personalized content for emails, newsletters, and social media posts. For example, if a donor has previously contributed to a specific project, these tools can generate personalized messages highlighting the impact of their contributions and inviting them to support similar initiatives. This approach not only acknowledges the donor's past support but also encourages continued engagement.

- **Automated Communication:**
  AI can automate communication processes, ensuring that donors receive timely and relevant information. For instance, a nonprofit might use **Chat with AI** to create automated responses for common donor inquiries, providing instant support and information. This prompt acknowledgment can enhance donor satisfaction and encourage future giving.

- **Chatbots for Engagement:**
  Implementing AI-powered chatbots, such as those offered by **Chat Arena**, on your website can provide instant support and information to potential donors. These chatbots can answer common questions, guide users through the donation process,

and even suggest donation amounts based on the user's previous giving history. By providing immediate assistance, chatbots can enhance the donor experience and increase conversion rates.

Predictive Analytics for Fundraising Success

Predictive analytics is a powerful tool that allows nonprofits to forecast future fundraising outcomes based on historical data and trends. By leveraging AI, organizations can make informed decisions about their fundraising strategies.

- **Forecasting Donations:**
  AI algorithms can analyze past donation patterns to predict future giving behavior. Tools like **AI Fact Checker** can help validate the data used in predictive models, ensuring that organizations base their forecasts on accurate information. For example, a nonprofit might use predictive analytics to identify trends in donor behavior during specific times of the year, such as holidays or awareness months. This information can help organizations plan targeted campaigns during peak giving periods, maximizing their fundraising potential.

- **Identifying Major Gift Prospects:**
  Predictive analytics can also help nonprofits identify potential major gift prospects. By using tools like **AI YouTube Summarizer**, organizations can analyze video content related to donor interests and engagement, highlighting individuals who are likely to make significant contributions. This insight allows organizations to focus their efforts on cultivating relationships with high-potential donors, increasing the likelihood of securing major gifts.

- **Optimizing Fundraising Strategies:**
  AI can provide recommendations for optimizing fundraising strategies based on predictive analytics. For instance, if data indicates that a particular donor segment responds well to matching gift campaigns, tools like **AI Text Humanizer** can help nonprofits tailor their outreach efforts accordingly. By aligning strategies with data-driven insights, organizations can enhance their fundraising effectiveness.

Case Studies of AI in Fundraising

Real-world examples of nonprofits successfully leveraging AI for fundraising can provide valuable insights and inspiration for organizations looking to enhance their own efforts.

- **Case Study 1: Charity: Water**
  Charity: Water, a nonprofit focused on providing clean drinking water to communities in need, has successfully utilized AI to enhance its fundraising efforts. By analyzing donor data with tools like **AI Image Generator** to create compelling visuals for campaigns, the organization identified key segments of its donor base and tailored its communication strategies accordingly. As a result, Charity: Water saw a significant increase in donor retention rates and overall contributions.

- **Case Study 2: The American Red Cross**
  The American Red Cross has implemented predictive analytics using **AI App Builder** to create custom applications that forecast donation trends and optimize its fundraising campaigns. By analyzing historical data, the organization identified patterns in donor behavior during disaster relief efforts. This insight

allowed the Red Cross to launch targeted campaigns during critical times, resulting in increased donations and support for its mission.

- **Case Study 3: World Wildlife Fund (WWF)**
  The WWF has leveraged AI to enhance its donor engagement strategies. By using AI-driven tools like **AI Voice Cloner** to create personalized voice messages for major donors, the organization has created communication that resonates with its supporters. This approach has led to increased donor satisfaction and higher rates of recurring contributions.

Question: How can you apply the SuperDonor Framework to deepen engagement with your donors?

Note Space: Identify 1-2 strategies to turn your donors into Super-Donors.

_____

_____

_____

_____

_____

_____

_____

_____

_____

_____

_____

_____

_____

_____

_____

_____

_____

_____

_____

CHAPTER 9

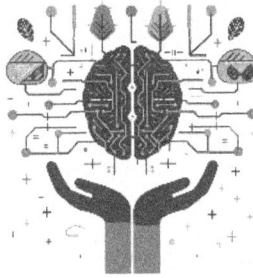

AI FOR GRANT WRITING
Streamlining Funding Success

Grant writing is the lifeblood of many nonprofits, providing the essential funding needed to sustain programs, expand services, and achieve mission-driven goals. However, the process is often time-consuming, resource-intensive, and fraught with challenges. From identifying the right funding opportunities to crafting compelling narratives and ensuring compliance with funder guidelines, grant writing demands precision, creativity, and strategic thinking.

Enter artificial intelligence (AI). AI has the potential to transform grant writing, making it faster, more efficient, and more effective. By automating repetitive tasks, generating high-quality content, and providing data-driven insights, AI can help nonprofits secure the funding they need to thrive.

We'll explore how nonprofits can leverage AI to streamline the grant writing process, from research and drafting to editing and submission. We'll also address the ethical considerations of using AI in grant writing and provide practical tips for integrating AI into your workflow.

Section 1: AI Tools for Grant Writing

AI offers a range of tools and applications that can simplify and enhance every stage of the grant writing process. Here's how nonprofits can use AI to their advantage:

1. Automated Research Tools

    Finding the right grant opportunities can feel like searching for a needle in a haystack. AI-powered research tools can save time by identifying funding opportunities that align with your nonprofit's mission and programs.

    - How It Works: These tools analyze your organization's profile, programs, and goals to match you with relevant grants.

    - Examples: Platforms like GrantStation, Instrumentl, and GrantWatch use AI to streamline the search process.

    - Benefit: Focus your efforts on grants with the highest likelihood of success.

2. Drafting and Editing Assistance

    Writing a grant proposal is both an art and a science. AI can help you generate high-quality drafts, refine your language, and ensure your proposal is clear, concise, and compelling.

    - How It Works: AI tools like ChatGPT, Jasper, and Grammarly can generate content, suggest improvements, and check for grammar and style errors.

- Benefit: Save time on drafting and editing while maintaining a professional tone.

3. Data Analysis and Impact Reporting

   Funders want to see measurable impact. AI can help you analyze program data and create compelling narratives that demonstrate your nonprofit's effectiveness.

   - How It Works: Tools like Tableau and Power BI visualize data, making it easier to communicate your impact in grant proposals.
   - Benefit: Strengthen your case with data-driven evidence.

4. Compliance and Funder Alignment

   Every funder has unique guidelines and priorities. AI can help ensure your proposal meets these requirements and aligns with the funder's goals.

   - How It Works: AI tools analyze funder guidelines and flag inconsistencies or missing elements in your proposal.
   - Benefit: Reduce the risk of rejection due to non-compliance.

Section 2: Ethical Considerations in AI Grant Writing

While AI offers significant benefits, it's essential to use it responsibly. Here are key ethical considerations to keep in mind:

1. Transparency and Authenticity

   AI-generated content should reflect your nonprofit's authentic voice and mission. Avoid over-reliance on AI to maintain the human touch in storytelling.

- Best Practice: Always review and personalize AI-generated content to ensure it aligns with your organization's values.

2. Data Privacy and Security

   Grant writing often involves sensitive information. Ensure that AI tools comply with data protection regulations (e.g., GDPR, CCPA) and safeguard your data.

   - Best Practice: Use trusted AI platforms with robust security measures.

3. Bias and Fairness

   AI tools can inadvertently introduce bias into your proposals. Be mindful of this and ensure your content is fair, inclusive, and representative of your community.

   - Best Practice: Review AI-generated content for potential biases and adjust as needed.

Section 3: Step-by-Step Guide to Using AI for Grant Writing

Here's a practical guide to integrating AI into your grant writing process:

1. Step 1: Research and Identify Opportunities

   Use AI tools to find grants that align with your nonprofit's mission and programs.

   - Action: Input your organization's profile into an AI-powered grant search tool and review the results.

2. Step 2: Draft the Proposal

   Use AI to generate an initial draft, incorporating key elements like the problem statement, objectives, and impact metrics.

- Action: Provide the AI tool with a clear prompt, such as, "Write a grant proposal for a program that supports underserved youth in our community."

3. Step 3: Refine and Personalize
   Edit the AI-generated content to reflect your nonprofit's unique voice and priorities. Add specific examples, anecdotes, and data to strengthen the proposal.
   - Action: Review the draft and make adjustments to ensure it aligns with your mission and funder's priorities.

4. Step 4: Review and Submit
   Use AI tools to check for compliance with funder guidelines and ensure the proposal is error-free.
   - Action: Run the final draft through an AI compliance checker and proofreading tool before submission.

Section 4: Case Studies of AI in Grant Writing

1. Case Study 1: Small Nonprofit Secures Major Grant
   A small nonprofit used AI to identify a funding opportunity and draft a winning proposal. By leveraging AI for research and content generation, they saved time and secured a $100,000 grant to expand their programs.

2. Case Study 2: Streamlining Grant Writing for a Large Organization
   A large nonprofit used AI to manage multiple grant applications simultaneously. By automating research, drafting, and compliance checks, they increased their success rate and reduced the workload for their grant writing team.

Section 5: Best Practices for AI-Driven Grant Writing

1. Start Small: Begin by using AI for specific tasks, like research or editing, before scaling up.
2. Collaborate with Your Team: Involve staff and stakeholders in the grant writing process to ensure AI-generated content aligns with your mission.
3. Continuously Improve: Use feedback from funders to refine your AI-driven grant writing strategies.
4. Stay Informed: Keep up with advancements in AI tools and best practices for grant writing.

AI has the potential to revolutionize grant writing, making it faster, more efficient, and more effective. By leveraging AI tools and strategies, nonprofits can secure the funding they need to achieve their missions and create lasting impact.

Question: What is one grant writing challenge your nonprofit faces that AI could help solve?

Note Space: Write down 1-2 steps you'll take to integrate AI into your grant writing process.

_____

_____

_____

_____

_____

_____

_____

_____

_____

_____

_____

_____

_____

_____

_____

_____

_____

Chapter 10

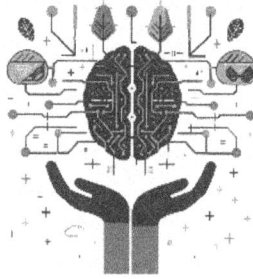

AI for Operations and Efficiency

If you have used CRM's, automated email responders or any system that has a progressive series of communication, you have done workflows and know it is tedious. AI will streamline workflows, improve operational efficiency, and ultimately enhance their ability to fulfill their missions. By leveraging AI tools and strategies, nonprofits can automate routine tasks, analyze data effectively, and optimize volunteer management, leading to more effective operations.

AI for Workflow Automation

Workflow automation is one of the most significant advantages of integrating AI into nonprofit operations. By automating repetitive tasks, organizations can free up valuable time and resources, allowing staff to focus on higher-impact activities.

- **Automating Administrative Tasks:**

  AI tools can handle various administrative functions, such as data entry and document management. For example, an **AI App Builder** can be used to create custom applications that automate data collection processes, such as gathering donor information from web forms and inputting it into a database. This automation reduces the risk of human error and ensures that data is consistently updated.

- **Streamlining Communication:**

  AI can also automate communication processes, such as sending reminders for upcoming events or follow-up emails to donors. An **AI Text to Speech** tool can be used to create audio messages for outreach, ensuring that supporters receive timely and relevant information in a format that is accessible to them. For instance, if a donor has not engaged with recent communications, the AI can trigger a personalized audio message to re-engage them.

- **Task Management:**

  AI tools can help nonprofits streamline their workflows by automating task assignments and tracking progress. An **AI Chat with PDF** tool can assist in managing project documentation by allowing team members to interact with project files directly, extracting relevant information quickly and efficiently. This capability ensures that everyone stays informed and accountable.

Data Analysis and Visualization with AI Tools

Data is a critical asset for nonprofits, and AI can enhance how organizations analyze and visualize this information. By leveraging AI tools, nonprofits can gain valuable insights that inform decision-making and strategy.

- **Data Analysis:**

  AI algorithms can process large datasets quickly and accurately, identifying trends and patterns that may not be immediately apparent. For example, an **AI Search Engine** can be used to analyze donor data and reveal insights about giving patterns, demographics, and engagement levels. This analysis can help nonprofits tailor their fundraising strategies and outreach efforts.

- **Data Visualization:**

  Effective data visualization is essential for communicating insights to stakeholders. An **AI Illustration Generator** can create visual representations of data, such as infographics or charts, that present information in a visually appealing and easily digestible format. For instance, a nonprofit might use this tool to create visual reports that showcase the impact of their programs, making it easier to share results with donors and board members.

- **Predictive Analytics:**

  AI can also be used for predictive analytics, allowing nonprofits to forecast future trends based on historical data. An **AI Content Detector** can analyze past communication and engagement data to predict when donors are likely to give again, helping organizations plan their fundraising campaigns accordingly. This proactive approach can lead to more effective resource allocation and improved fundraising outcomes.

AI for Volunteer Recruitment and Scheduling

Volunteers are a vital resource for many nonprofits, and AI can stream-line the recruitment and scheduling processes to ensure that organizations can effectively mobilize their volunteer workforce.

- **Recruitment Automation:**
  AI tools can assist in identifying potential volunteers by analyzing social media profiles and community engagement. An **AI Chat Arena** can be used to interact with potential volunteers on social media platforms, answering questions and guiding them to sign up for opportunities that align with their skills and interests, making the recruitment process more efficient.

- **Scheduling and Coordination:**
  AI can simplify volunteer scheduling by automating the process of assigning shifts and managing availability. An **AI Voice Cloner** can be used to create personalized voice messages for volunteers, reminding them of their upcoming shifts and providing important information. This capability can reduce no-shows and improve overall attendance.

- **Engagement and Retention:**
  AI can also enhance volunteer engagement by providing personalized communication and recognition. An **AI Text Humanizer** can be used to craft personalized thank-you messages to volunteers after their shifts, highlighting the impact of their contributions. This acknowledgment fosters a sense of belonging and encourages volunteers to remain engaged with the organization.

Case Studies of AI in Nonprofit Operations

Real-world examples of nonprofits successfully leveraging AI for operational efficiency can provide valuable insights and inspiration for organizations looking to enhance their own efforts.

- **Case Study 1: Habitat for Humanity**

  Habitat for Humanity has implemented AI tools to streamline its volunteer management processes. By using an **AI App Builder**, the organization has automated volunteer recruitment and scheduling, allowing staff to focus on building homes rather than managing logistics. This efficiency has led to increased volunteer participation and improved project outcomes.

- **Case Study 2: The Nature Conservancy**

  The Nature Conservancy has utilized AI-driven data analysis tools to enhance its conservation efforts. By analyzing environmental data with an **AI Search Engine**, the organization has gained insights into the effectiveness of its programs and initiatives. This data-driven approach has allowed the Nature Conservancy to allocate resources more effectively and maximize its impact on conservation efforts.

- **Case Study 3: Feeding America**

  Feeding America has leveraged AI to optimize its food distribution network. By using predictive analytics with an **AI Content Detector**, the organization can forecast food demand in different regions and adjust its supply chain accordingly. This optimization has led to more efficient food distribution, ensuring that resources reach those in need more effectively.

AI has the potential to transform nonprofit operations by streamlining workflows, enhancing data analysis, and improving volunteer management. By leveraging AI tools and strategies, organizations can increase their operational efficiency, allowing them to focus on their core mission and achieve greater impact in their communities. As technology continues to evolve, the role of AI in nonprofit operations will likely expand, offering even more opportunities for organizations to enhance their effectiveness and reach.

Question: What volunteer management challenges could AI help solve in your organization?

Note Space: Write down 1-2 ways AI could streamline your volunteer recruitment or scheduling process.

_____

_____

_____

_____

_____

_____

_____

_____

_____

_____

_____

_____

_____

_____

_____

_____

_____

_____

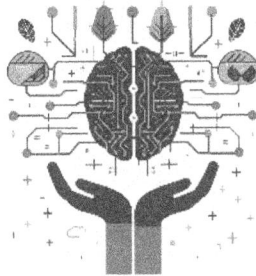

Chapter 11
AI for Storytelling and Impact Reporting

Storytelling is a vital tool for nonprofits, as it engages supporters, communicates the mission, and demonstrates the organization's impact. By leveraging AI tools, nonprofits can enhance their storytelling capabilities, visualize data effectively, and automate impact reporting workflows, ultimately leading to more effective communication and engagement.

AI Tools for Storytelling

Storytelling is at the heart of nonprofit communication, and AI can significantly enhance this process by providing tools that help organizations create engaging narratives.

- **AI Content Generators:**

  Tools like **Custom GPTs** can assist nonprofits in generating compelling narratives based on specific prompts. For example, a nonprofit focused on animal rescue can input details about a recent rescue story, and the AI can generate a heartfelt narrative that highlights the animal's journey and the organization's role in its recovery. This capability allows organizations to produce high-quality content quickly, ensuring that stories are shared in a timely manner.

- **AI Image and Video Generators:**
  Visual storytelling is essential for capturing attention and conveying emotions. AI tools such as **AI Image Generators** and **AI Video Generators** can create visuals that complement written narratives. For instance, a nonprofit can use an AI Image Generator to create illustrations that depict the impact of its programs, while an AI Video Generator can produce short videos that showcase success stories, making the content more engaging and shareable.

- **AI Music Generators:**
  Incorporating music into storytelling can enhance emotional resonance. An **AI Music Generator** can create original soundtracks for videos or presentations, setting the tone for the story being told. For example, a nonprofit might use uplifting music to accompany a video that highlights the positive outcomes of its programs, further engaging viewers and evoking emotional responses.

Data Visualization with AI

Data visualization is a powerful way to communicate impact, and AI can enhance how nonprofits present their data to stakeholders.

- **Creating Interactive Dashboards:**
  AI tools like **AI Illustration Generators** can help nonprofits

create visually appealing infographics and dashboards that present data in an easily digestible format. For example, a nonprofit focused on education can visualize statistics related to student performance, program reach, and funding sources, making it easier for stakeholders to understand the organization's impact at a glance.

- **Dynamic Data Representation:**
AI can also automate the process of updating visualizations based on real-time data. For instance, an **AI Search Engine** can pull the latest statistics from various sources and automatically update visual reports, ensuring that stakeholders always have access to the most current information. This dynamic representation of data can enhance transparency and build trust with donors and supporters.

- **Storytelling with Data:**
Combining storytelling with data visualization can create a compelling narrative that highlights the organization's impact. For example, a nonprofit can use an **AI Content Detector** to analyze donor feedback and identify key themes, then incorporate these insights into a visual report that tells the story of how donor contributions have made a difference. This approach not only informs stakeholders but also engages them emotionally.

Automating Impact Reporting Workflows

Impact reporting is essential for demonstrating accountability and transparency, and AI can streamline this process.

- **Automated Data Collection:**
AI tools can automate the collection of data needed for impact

reports. For instance, an **AI Transcription** tool can convert audio recordings of interviews or focus groups into written text, making it easier to gather qualitative data for reports. This automation saves time and ensures that important insights are not overlooked.

- **Streamlining Report Generation:**
  AI can assist in generating impact reports by compiling data, visualizations, and narratives into a cohesive document. Tools like **Chat with PDF** can help organizations create interactive reports that allow stakeholders to explore data and insights in a user-friendly format. For example, a nonprofit can generate a report that includes clickable sections for different programs, allowing readers to dive deeper into specific areas of interest.

- **Personalized Reporting:**
  AI can also help tailor impact reports to different audiences. By analyzing stakeholder preferences and engagement history, AI tools can suggest which data points and stories to highlight for specific donors or partners. This personalized approach can enhance the relevance of reports and strengthen relationships with supporters.

Examples of How to Leverage Stories

Leveraging stories effectively can amplify a nonprofit's message and engage supporters on a deeper level. Here are some examples of how organizations can use storytelling to their advantage:

- **Highlighting Individual Impact:**
  Sharing individual success stories can create a powerful emotional connection with supporters. For instance, a nonprofit that

provides scholarships can share the story of a student who over-came challenges to achieve academic success, illustrating the direct impact of donor contributions.

- **Showcasing Community Change:**
Nonprofits can tell stories that highlight broader community impact. For example, a food bank can share narratives about families who have benefited from its services, showcasing how the organization is addressing food insecurity in the community. This approach not only informs donors but also inspires them to support ongoing efforts.

- **Engaging Through Social Media:**
Social media platforms are ideal for sharing stories in real-time. Nonprofits can use AI tools to create engaging posts that combine visuals, narratives, and data. For example, a wildlife conservation organization can share a series of posts that document the journey of a rehabilitated animal, using AI-generated images and videos to enhance the storytelling experience.

- **Creating Compelling Campaigns:**
Nonprofits can develop campaigns centered around a specific story or theme. For instance, a health organization might launch a campaign focused on mental health awareness, sharing personal stories from individuals who have benefited from its programs. This approach can mobilize supporters and encourage them to take action, whether through donations, volunteering, or advocacy.

By leveraging AI for content generation, data visualization, and workflow automation, organizations can craft compelling narratives that engage supporters and effectively communicate their impact. As non-profits continue to embrace AI technology, they can create more meaningful connections with their audiences, ultimately driving greater support for their missions.

Question: What impact metrics are most important to your stakeholders, and how could AI help you track and report them?

Note Space: List 3 key metrics you'd like to measure and visualize using AI tools.

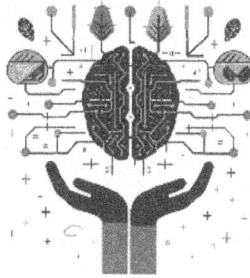

Chapter 12
AI for Impact Measurement and Reporting

Tracking measurements is essential for nonprofits to understand the effectiveness of their programs and initiatives. This chapter will enable nonprofits to effectively track outcomes and showcase their impact using artificial intelligence (AI). By leveraging AI tools and strategies, organizations can gather and analyze data, automate reporting processes, and visualize their impact in compelling ways. Understanding and communicating impact is crucial for building trust with stakeholders, securing funding, and driving community engagement.

The Importance of Tracking Measurements

By collecting and analyzing data, organizations can gain insights into their performance, identify areas for improvement, and demonstrate their impact to stakeholders. Here are several reasons why tracking measurements is vital:

- **Demonstrating Accountability:**

  Nonprofits are often accountable to donors, grantors, and the communities they serve. By tracking outcomes, organizations can provide evidence of their effectiveness and show how funds are being utilized. This transparency builds trust and credibility, encouraging continued support from stakeholders.

- **Informed Decision-Making:**

  Data-driven decision-making is crucial for optimizing programs and strategies. By analyzing performance metrics, nonprofits can identify which initiatives are successful and which may need adjustments. For example, if a program aimed at reducing homelessness shows low success rates, the organization can investigate the underlying causes and make necessary changes to improve outcomes.

- **Identifying Trends and Patterns:**

  Tracking measurements allows nonprofits to identify trends and patterns over time. This information can help organizations anticipate future needs and adapt their strategies accordingly. For instance, if data shows an increase in demand for mental health services, a nonprofit can allocate resources to expand its offerings in that area.

- **Enhancing Fundraising Efforts:**

  Impact data can be a powerful tool for fundraising. By showcasing the results of their work, nonprofits can create compelling narratives that resonate with potential donors. For example, a nonprofit that tracks the number of individuals served and the outcomes achieved can use this data to craft impactful stories that illustrate the difference their work makes in the community.

- **Improving Program Design:**
  Continuous measurement and evaluation enable nonprofits to refine their programs based on real-world data. By understanding what works and what doesn't, organizations can design more effective interventions that better meet the needs of their target populations.

AI Tools for Impact Measurement

AI tools can significantly enhance the process of impact measurement by automating data collection, analysis, and reporting. Here are some key AI tools that nonprofits can leverage:

- **Tableau:**
  Powered by Salesforce, Tableau is a powerful data visualization tool that allows nonprofits to create interactive dashboards and reports. By integrating data from various sources, organizations can visualize their impact in real-time, making it easier to communicate results to stakeholders. For example, a nonprofit focused on education can use Tableau to track student performance metrics and visualize progress over time.

- **Power BI:**
  Microsoft Power BI is another robust analytics tool that enables nonprofits to analyze and visualize data. With its user-friendly interface, organizations can create custom reports that highlight key performance indicators (KPIs) and outcomes. For instance, a nonprofit working in healthcare can use Power BI to track patient outcomes and visualize trends in service delivery.

- **Google Analytics:**

  While primarily used for website analytics, Google Analytics can also provide valuable insights into user engagement and program effectiveness. Nonprofits can track how visitors interact with their website, which can inform outreach strategies and improve user experience.

Automating Grant Writing and Reporting with AI

AI can streamline the grant writing and reporting processes, saving nonprofits time and resources while improving the quality of their submissions.

- **Automated Grant Writing Tools:**

  AI-driven grant writing tools can assist nonprofits in drafting proposals by providing templates, suggestions, and data-driven insights. For example, tools like **GrantWriterAI** can analyze successful grant applications and generate tailored proposals based on the organization's mission and goals. This automation can increase the chances of securing funding by ensuring that proposals are well-structured and aligned with funder priorities.

- **Streamlined Reporting Processes:**

  AI can automate the process of generating impact reports by compiling data, visualizations, and narratives into cohesive documents. For instance, an **AI Transcription** tool can convert meeting notes and discussions into written reports, ensuring that important insights are captured and included in the final document. This efficiency allows nonprofits to focus on analyzing results rather than spending excessive time on report preparation.

Visualizing Impact Data for Storytelling

Visualizing impact data is crucial for effective storytelling. By presenting data in a visually appealing and easily digestible format, nonprofits can engage stakeholders and communicate their impact more effectively.

- **Creating Compelling Visuals:**
  AI tools like **Tableau** and **Power BI** can help nonprofits create infographics, charts, and interactive dashboards that showcase their impact. For example, a nonprofit focused on environmental conservation can use visualizations to illustrate the number of trees planted, areas restored, and wildlife protected, making the data more relatable and impactful.

- **Integrating Stories with Data:**
  Combining qualitative stories with quantitative data can create a powerful narrative that resonates with audiences. For instance, a nonprofit can share a success story of an individual who benefited from its programs alongside data that highlights the overall impact of those programs. This approach not only informs stakeholders but also evokes emotional responses that drive engagement.

Best Practices for AI-Driven Reporting

To maximize the effectiveness of AI-driven reporting, nonprofits should consider the following best practices:

- **Define Clear Metrics:**
  Establish clear and measurable outcomes that align with the organization's goals. This clarity will guide data collection and analysis efforts, ensuring that the right information is captured.

- **Utilize Multiple Data Sources:**
  Integrate data from various sources to gain a comprehensive

understanding of impact. This can include quantitative data (e.g., program statistics) and qualitative data (e.g., participant testimonials) to provide a well-rounded view of outcomes.

- **Regularly Review and Update Reports:**
  Impact reporting should be an ongoing process. Regularly review and update reports to reflect the most current data and insights. This practice ensures that stakeholders receive timely information and demonstrates the organization's commitment to transparency.

- **Engage Stakeholders in the Process:**
  Involve stakeholders in the measurement and reporting process. This engagement can provide valuable feedback and insights that enhance the quality of reports and strengthen relationships with supporters.

AI offers powerful tools and strategies that can enhance impact measurement and reporting for nonprofits. By leveraging AI for data collection, analysis, and visualization, organizations can track outcomes effectively and showcase their impact to stakeholders. Understanding the importance of tracking measurements and utilizing data-driven insights will empower nonprofits to make informed decisions, improve their programs, and ultimately achieve greater impact in their communities.

Question: How could you use data visualization to tell a compelling story about your nonprofit's impact?

Note Space: Sketch out a concept for an infographic or dashboard that showcases your impact.

_____

_____

_____

_____

_____

_____

_____

_____

_____

_____

_____

_____

_____

_____

_____

_____

_____

_____

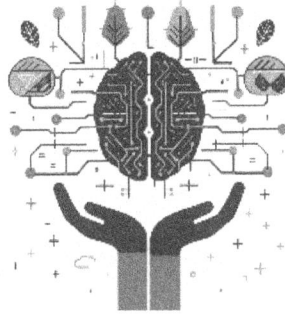

PART 3: STRATEGIC PLANNING WITH AI

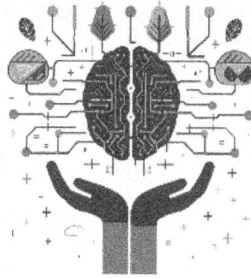

Chapter 13
Logic Model Framework and AI

Logic Model Framework serves as a foundational tool for nonprofits, providing a clear framework for program planning and evaluation. The Philantrepreneur Foundation believes that programs are the heartbeat of organizations and what drives their mission. By effectively utilizing Logic Models, nonprofits can articulate their goals, identify the resources needed, and outline the expected outcomes of their programs. This clarity is crucial for aligning team efforts and securing stakeholder buy-in.

Tracking measurements through Logic Models is essential for nonprofits to understand the effectiveness of their programs and initiatives. By collecting and analyzing data, organizations can gain insights into their performance, identify areas for improvement, and demonstrate their impact to stakeholders. Here's why Logic Models are vital:

- Clarifying Program Goals:
  Logic Models help nonprofits define their program goals and objectives clearly. By mapping out the inputs, activities, outputs, and outcomes, organizations can ensure that everyone involved understands the purpose of the program and what it aims to achieve.

- Connecting Resources to Outcomes:
  A well-constructed Logic Model illustrates the relationship between resources (inputs), activities, and desired outcomes. This connection helps nonprofits understand how their investments translate into impact. For example, if a nonprofit invests in training staff (input), conducts workshops (activity), and aims to improve community literacy (outcome), the Logic Model visually represents this flow, making it easier to communicate to stakeholders.

- Facilitating Evaluation and Learning:
  Logic Models provide a framework for evaluating program effectiveness. By clearly defining expected outcomes, nonprofits can measure their success against these benchmarks. This evaluation process not only helps organizations assess their impact but also informs future program improvements.

- Enhancing Communication:
  Logic Models serve as powerful communication tools. They can be used to convey program strategies and expected outcomes to stakeholders, including funders, board members, and community partners. A well-designed Logic Model can succinctly

illustrate what a program is doing and the impact it aims to achieve, fostering transparency and trust.

- Key Function Before Strategic Planning:
Completing a Logic Model is a key function or process that should be undertaken before starting strategic planning. The data output generated from the Logic Model is essential for informing the strategic planning process, ensuring that decisions are based on evidence and aligned with the organization's mission.

Visual Map

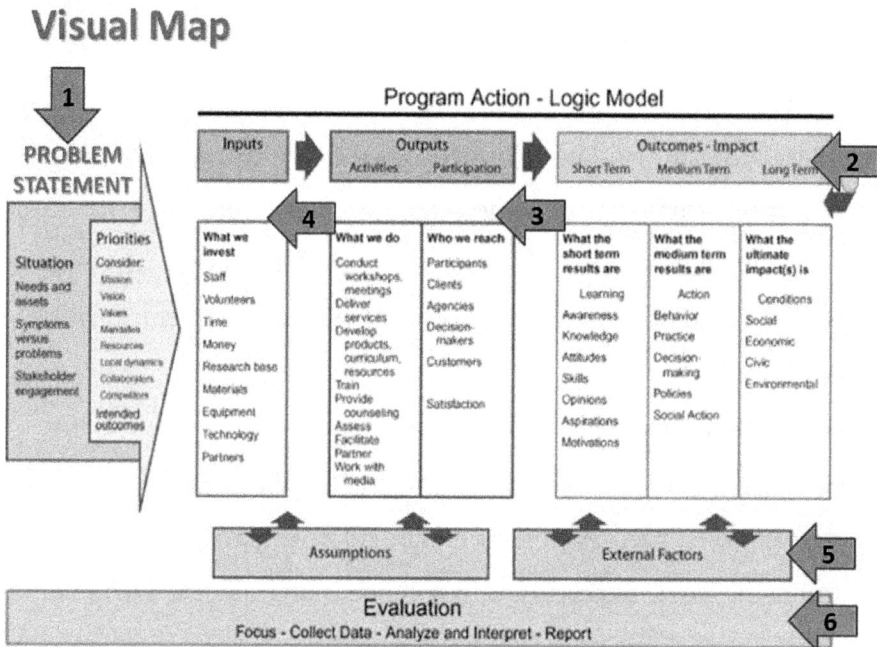

Program Action - Logic Model

1 PROBLEM STATEMENT	Inputs	Outputs Activities Participation	Outcomes - Impact Short Term Medium Term Long Term **2**

Priorities	What we invest	What we do	Who we reach	What the short term results are	What the medium term results are	What the ultimate impact(s) is
Situation	Consider:	Staff	Conduct workshops, meetings	Participants	Learning	Action	Conditions
Needs and assets	Mission	Volunteers	Deliver services	Clients	Awareness	Behavior	Social
Symptoms versus problems	Vision Values Mandates	Time	Develop products, curriculum, resources	Agencies	Knowledge	Practice	Economic
Stakeholder engagement	Resources Local dynamics	Money	Train	Decision-makers	Attitudes	Decision-making	Civic
	Collaborators Competitors	Research base	Provide counseling	Customers	Skills	Policies	Environmental
	Intended outcomes	Materials	Assess	Satisfaction	Opinions	Social Action	
		Equipment	Facilitate		Aspirations		
		Technology	Partner		Motivations		
		Partners	Work with media				

Assumptions	External Factors **5**

Evaluation
Focus - Collect Data - Analyze and Interpret - Report **6**

If you need assistance with Logic Model development, The Philantrepreneur Foundation offers individual and team training to help organizations effectively implement this framework.

Integrating AI into the Logic Model Framework

AI can enhance the Logic Model framework by providing tools for data collection, analysis, and optimization. Here's how nonprofits can leverage AI in their program planning processes:

- AI Tools for Data Collection and Analysis:

  AI-driven tools like Tableau and Power BI can help nonprofits collect and analyze data related to their programs. By integrating these tools into the Logic Model framework, organizations can track key performance indicators (KPIs) and visualize their impact in real-time. For example, a nonprofit focused on health services can use these tools to analyze patient outcomes and adjust their programs based on data insights.

- Using AI to Predict Program Outcomes:

  AI can assist nonprofits in predicting program outcomes based on historical data. By analyzing past performance metrics, AI algorithms can identify trends and forecast future results. This predictive capability allows organizations to make informed decisions about resource allocation and program adjustments. For instance, if data indicates that a particular intervention is consistently effective, nonprofits can allocate more resources to that program.

- Optimizing Resource Allocation:

  AI can help nonprofits optimize their resource allocation by analyzing data on program effectiveness and community needs. By using AI tools to assess which programs yield the highest impact, organizations can prioritize their investments and ensure that resources are directed where they are most needed. This strategic approach enhances overall program effectiveness and sustainability.

Visualizing Impact Data for Storytelling

Visualizing impact data is crucial for effective storytelling. By presenting data in a visually appealing and easily digestible format, nonprofits can engage stakeholders and communicate their impact more effectively.

- Creating Compelling Visuals:
  AI tools like Tableau and Power BI can help nonprofits create infographics, charts, and interactive dashboards that showcase their impact. For example, a nonprofit focused on environmental conservation can use visualizations to illustrate the number of trees planted, areas restored, and wildlife protected, making the data more relatable and impactful.

- Integrating Stories with Data:
  Combining qualitative stories with quantitative data can create a powerful narrative that resonates with audiences. For instance, a nonprofit can share a success story of an individual who benefited from its programs alongside data that highlights the overall impact of those programs. This approach not only informs stakeholders but also evokes emotional responses that drive engagement.

Best Practices for AI-Driven Logic Models

To maximize the effectiveness of AI-driven Logic Models, nonprofits should consider the following best practices:

- Define Clear Metrics:
  Establish clear and measurable outcomes that align with the

organization's goals. This clarity will guide data collection and analysis efforts, ensuring that the right information is captured.

- Utilize Multiple Data Sources:
  Integrate data from various sources to gain a comprehensive understanding of impact. This can include quantitative data (e.g., program statistics) and qualitative data (e.g., participant testimonials) to provide a well-rounded view of outcomes.

- Regularly Review and Update Logic Models:
  Logic Models should be living documents that are regularly reviewed and updated to reflect the most current data and insights. This practice ensures that stakeholders receive timely information and demonstrates the organization's commitment to transparency.

- Engage Stakeholders in the Process:
  Involve stakeholders in the Logic Model development process. This engagement can provide valuable feedback and insights that enhance the quality of the Logic Model and strengthen relationships with supporters.

Integrating AI into the Logic Model framework can significantly enhance program planning for nonprofits. By leveraging AI tools for data collection, analysis, and prediction, organizations can optimize their programs and demonstrate their impact effectively. Understanding the importance of Logic Models and the role of AI in program planning will empower nonprofits to make informed decisions, improve their programs, and ultimately achieve greater impact in their communities.

Question: How could AI enhance your nonprofit's program planning and evaluation process?

Note Space: Identify 1-2 programs where AI could improve data collection or analysis.

_____

_____

_____

_____

_____

_____

_____

_____

_____

_____

_____

_____

_____

_____

_____

_____

_____

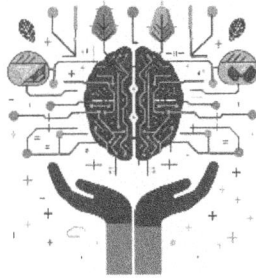

Chapter 14
AI for Strategic Planning

In the dynamic landscape of nonprofit organizations, effective strategic planning is essential for achieving missions and maximizing impact. A well-defined strategic plan serves as a roadmap, guiding organizations in their efforts to address community needs and allocate resources efficiently. Strategic planning is not a one-time event but an ongoing process that requires continuous reflection and adaptation. It is essential for nonprofits to recognize that their strategic plans should always be a work-in-progress. As the environment in which they operate changes— whether due to shifts in community needs, funding landscapes, or organizational capacity, so too must their strategies.

By integrating artificial intelligence (AI) into the strategic planning process, nonprofits can enhance their ability to gather and analyze data, predict outcomes, and optimize resource allocation. AI tools can provide valuable insights that inform decision-making, allowing organizations to adapt their strategies based on real-time data and trends.

Phases of Strategic Planning

1. **Innovative Strategic Thinking:**

   The first phase of strategic planning involves fostering a culture of innovative thinking within the organization. Nonprofits should encourage team members to think outside the box and explore new concepts that can enhance their programs and services. This phase is crucial for breaking free from traditional approaches and embracing creative solutions that can lead to greater impact.

AI can play a significant role in this phase by providing tools that facilitate brainstorming and idea generation. For example, AI-driven platforms can analyze trends in the nonprofit sector, helping organizations identify emerging opportunities and challenges. By leveraging these insights, nonprofits can develop innovative strategies that align with their mission and community needs.

2. **Strategic Planning:**

   Once innovative ideas have been generated, the next phase involves developing a comprehensive strategic plan. This plan should outline the organization's goals, objectives, and the strategies needed to achieve them.

AI tools can assist in this phase by automating data collection and analysis, providing insights that inform the strategic planning process. For instance, AI can analyze historical performance data to identify which programs have been most effective, allowing organizations to prioritize their efforts accordingly.

3. **Operational Planning:**

   After the strategic plan is developed, the focus shifts to operational planning. This phase involves translating the strategic goals into actionable steps and allocating resources effectively. AI can support this process by optimizing resource allocation based on data-driven insights. For example, AI algorithms can analyze community needs and program effectiveness to recommend where to direct funding and staff resources.

4. **Measurement Planning:**

   The final phase of strategic planning involves establishing metrics for measuring success. This phase is critical for evaluating the effectiveness of programs and ensuring accountability to stakeholders. AI tools can enhance measurement planning by automating data collection and analysis, providing real-time insights into program performance.

By integrating AI into the measurement planning process, nonprofits can track key performance indicators (KPIs) and assess their impact more effectively. This data-driven approach allows organizations to make informed decisions about program adjustments and improvements.

Using AI to Develop and Maintain Strategic Plans

AI can be a powerful ally in both the development and maintenance of strategic plans. Here are some ways nonprofits can leverage AI throughout the strategic planning process:

- **Data-Driven Insights:**

  AI tools can analyze vast amounts of data to provide insights that inform strategic decisions. By leveraging AI for data analysis, nonprofits can identify trends, assess community needs, and evaluate program effectiveness, ensuring that their strategic plans are grounded in evidence.

- **Continuous Monitoring and Adaptation:**
  Strategic plans should be dynamic documents that evolve over time. AI can facilitate continuous monitoring of program performance and community needs, allowing organizations to adapt their strategies as necessary. For example, AI-driven dashboards can provide real-time updates on key metrics, enabling nonprofits to respond quickly to changing circumstances.

- **Engaging Stakeholders:**
  AI can enhance stakeholder engagement by providing personalized communication and updates on strategic initiatives. By using AI tools to analyze stakeholder preferences and behaviors, nonprofits can tailor their outreach efforts, ensuring that stakeholders remain informed and engaged in the strategic planning process.

Understanding the importance of strategic planning and the role of AI in this process will empower nonprofits to make informed decisions, improve their programs, and ultimately achieve greater success in their communities. It can significantly enhance the effectiveness of nonprofit organizations by fostering innovative strategic thinking, utilizing AI tools for data collection and analysis, and continuously adapting strategies based on real-time insights, nonprofits can create strategic plans that drive their mission and create meaningful impact.

Question: What strategic goal could AI help your nonprofit achieve in the next 6-12 months?

Note Space: Outline a plan to integrate AI into your strategic planning process.

Chapter 15

The Future of AI in Nonprofits – Adapt to Survive, Adapt to Thrive

In an era marked by rapid technological advancements, the nonprofit sector faces both challenges and opportunities. The integration of artificial intelligence (AI) into organizational strategies is no longer a luxury but a necessity for nonprofits aiming to thrive in a competitive landscape. As technology continues to evolve at an unprecedented pace, it is critical for organizations to keep up with these advancements to remain relevant and effective in their missions.

The Rapid Advancement of Technology

Over the past few years, we have witnessed a remarkable acceleration in technological innovation. From the rise of machine learning and natural language processing to the proliferation of data analytics tools, AI has transformed how organizations operate and engage with their stakeholders. For instance, just a decade ago, many nonprofits were still relying on traditional methods for data collection and analysis. Today, AI tools can process vast amounts of data in real-time, providing insights that were previously unimaginable. As a result, those that embraced technology not only survived but thrived, demonstrating the importance of adaptability in the face of change.

Emerging Trends in AI for Nonprofits

As we look to the future, several emerging trends in AI are poised to shape the nonprofit landscape:

- **Enhanced Data Analytics:**
  AI-driven analytics tools will continue to evolve, enabling nonprofits to gain deeper insights into donor behavior, program effectiveness, and community needs. By harnessing these insights, organizations can make data-informed decisions that enhance their impact.

- **Personalization and Engagement:**
  AI will play a crucial role in personalizing communication and engagement strategies. By analyzing user data, nonprofits can tailor their messaging to resonate with individual supporters, fostering stronger connections and increasing donor retention.

- **Automation of Routine Tasks:**
  The automation of administrative tasks through AI will free up valuable time for nonprofit staff, allowing them to focus on mission-driven activities. From automating donor communications to streamlining grant writing processes, AI will enhance operational efficiency.

- **Predictive Analytics for Program Planning:**
  AI will enable nonprofits to leverage predictive analytics to forecast program outcomes and optimize resource allocation. By understanding potential future trends, organizations can proac-
  tively address community needs and enhance their service delivery.

How to Stay Updated on AI Developments

To thrive in this rapidly changing environment, nonprofits must prioritize staying informed about AI developments. The Nonprofit AI Academy—a comprehensive suite of resources, training, and tools—makes it
easy for organizations to keep up with emerging trends. Here's how you
can stay ahead:

1. **Engage with AI Communities Through Membership**
   Join the **Nonprofit AI Academy Membership**, a platform
   designed to connect you with experts, peers, and thought leaders
   in the nonprofit AI space. Gain access to exclusive webinars, fo-
   rums, and conferences that provide valuable insights, network-
   ing opportunities, and real-world case studies. Stay informed
   about the latest advancements and best practices while building
   a community of like-minded professionals.

2. **Invest in Training with the Accelerated AI Course**
   Continuous learning is essential for adapting to new technologies. The Accelerated AI Course, part of the Nonprofit AI Academy, equips your team with the knowledge and skills to effectively implement AI in your organization. This hands-on training program covers everything from AI basics to advanced

applications, ensuring your nonprofit is prepared to leverage AI responsibly and strategically.

3. **Leverage Tools and Resources with the AI Toolkit**

   The Nonprofit AI Toolkit, included in the Nonprofit AI Academy, provides curated resources, templates, and tools to help you stay ahead of the curve. From ethical AI frameworks to step-by-step implementation guides, the toolkit is your go-to resource for practical, actionable insights. Regularly updated with the latest trends and expert advice, it ensures your organization remains at the forefront of AI innovation.

By leveraging the **Nonprofit AI Academy's** membership, Accelerated AI Course, and AI Toolkit, your organization can stay informed, build capacity, and confidently navigate the evolving AI landscape. These resources are designed to help you harness the power of AI while staying true to your mission and values.

Building a Culture of Innovation and Adaptability

Creating a culture of innovation and adaptability is crucial for nonprofits looking to embrace AI as a long-term strategy. Here are some key practices to foster this culture:

- **Encourage Experimentation:**
  Organizations should create an environment where team members feel empowered to experiment with new ideas and technologies. Encouraging a mindset of curiosity and exploration can lead to innovative solutions that enhance program effectiveness.

- **Foster Collaboration:**
  Collaboration across departments and with external partners

can drive innovation. By sharing knowledge and resources, non-profits can leverage diverse perspectives to develop creative solutions that address community needs.

- **Embrace Change:**
  Nonprofits must be willing to embrace change and adapt their strategies as needed. This flexibility is essential for navigating the evolving landscape of technology and ensuring that organizations remain relevant and effective.

As we look to the future, AI will play a transformative role in the nonprofit sector. Organizations that embrace AI as a long-term strategy will be better positioned to adapt to changing circumstances, enhance their impact, and drive meaningful change in their communities.

The Philantrepreneur Foundation is committed to supporting nonprofits on this journey. We encourage organizations to take advantage of our training resources, workshops, and membership platforms to build their capacity in AI and stay informed about emerging trends. Together, we can navigate the future of AI in nonprofits and create a lasting impact.

NEXT STEP - AI Implementation Checklist for Nonprofits

Use these 10 steps to ensure you have covered all your bases.

1. Define Your Objectives

- **Identify Goals**: What specific outcomes do you want to achieve with AI? (e.g., increase donor engagement, streamline operations)

- **Align with Mission**: Ensure that your AI objectives align with your organization's overall mission and strategic goals.

2. Assess Current Capabilities

- **Evaluate Existing Resources**: What technology, data, and human resources do you currently have?

- **Identify Gaps**: What skills or tools are missing that are necessary for successful AI implementation?

3. Understand Your Data

- **Data Inventory**: What data do you currently collect? Is it clean, accurate, and relevant?

- **Data Privacy**: Ensure compliance with data protection regulations (e.g., GDPR, CCPA) and establish a plan for data security.

4. Research AI Tools

- **Identify Potential Tools**: What AI tools are available that align with your objectives? (e.g., predictive analytics, chatbots)

- **Evaluate Suitability**: Assess the features, costs, and user-friendliness of each tool.

5. Develop a Strategy

- **Create an Implementation Plan**: Outline the steps needed to integrate AI tools, including timelines and responsibilities.

- **Set Key Performance Indicators (KPIs)**: Define how you will measure success (e.g., increased donations, improved efficiency).

6. Engage Stakeholders

- **Involve Team Members**: Ensure that staff members are informed and involved in the AI adoption process.

- **Communicate Benefits**: Clearly articulate how AI will benefit the organization and its mission to gain buy-in from stakeholders.

7. Pilot Testing

- **Conduct a Pilot Program**: Test the AI tool on a small scale to evaluate its effectiveness and gather feedback.

- **Monitor Performance**: Track the results of the pilot against your KPIs to assess success.

8. Training and Support

- **Provide Training**: Ensure that staff members receive adequate training on how to use the AI tools effectively.

- **Establish Support Systems**: Create channels for ongoing support and troubleshooting as staff adapt to the new tools.

9. Review and Adjust

- **Evaluate Outcomes**: After implementation, review the results against your objectives and KPIs.

- **Make Adjustments**: Be prepared to refine your approach based on feedback and performance data.

10. Scale and Expand

- **Plan for Scaling**: If the pilot is successful, develop a plan to scale the AI implementation across the organization.

- **Explore Additional Applications**: Consider other areas where AI could further enhance your operations and impact.

Question: What steps will you take to foster a culture of innovation and adaptability in your nonprofit?

Note Space: Write down 1-2 actions you'll implement to embrace AI as a long-term strategy.

RESORCES

The Nonprofit AI Academy: Your Long-Term Solution for AI Mastery

The **Nonprofit AI Academy** is your gateway to staying ahead in an ever-evolving digital landscape. As AI continues to reshape the way nonprofits operate, fundraise, and engage with their communities, ongoing learning is essential, not just to keep up, but to thrive. The Academy is designed to be your long-term solution, providing the continuous education, resources, and support needed to navigate and leverage AI effectively.

Enjoy a **FREE** membership to get access to the basics, or level up to unlock exclusive training, advanced tools, and expert insights tailored to your nonprofit's needs. With multiple membership tiers designed to meet every level of AI adoption, the Academy provides the knowledge and tools to help you stay ahead and drive lasting impact.

Learn more about the NONPROFIT AI ACADEMY

visit: NonprofitAIAcademy.org

Nonprofit AI Readiness Scorecard Quiz

Simple quiz **https://PhilantrepreneurFoundation.org/AIScorecard**

https://www.youtube.com/@BestAIforNonprofits

https://www.youtube.com/@TPFNetwork

Nonprofit Corner Podcast

Dr. Victoria Boyd - https://DrVictoriaBoyd.com

NONPROFIT AI PLAYBOOK TOOLKIT!

https://tinyurl.com/5zpmufjd

Make this toolkit your 'go-to' resource. It contains guides, checklists, and templates. New material will be added regularly. For direct access use this link or QR code.

Galaxy.ai Galaxy.ai is a game changer in the world of AI tools and platforms, offering a unique integration of multiple technologies that eliminates the need for multiple tools and their associated cost. Signup today, use the QR code or **https://DrVictoriaBoyd.com/GALAXY**.

QRCODECHIMP
Create static and dynamic QR codes

TOOLS: *Please note we do not endorse or recommend specifically any of these tools. They are for your reference.*

Fundraising & Donor Engagement
1. **DonorSearch AI** – Uses predictive analytics to identify potential major donors based on wealth and philanthropic history.
2. **Graystone AI** – Helps optimize fundraising campaigns by analyzing donor behavior and suggesting engagement strategies.
3. **ThankView** – AI-powered personalized video messaging to enhance donor stewardship.

Grant Writing & Research
4. **Grantable** – AI-driven grant writing assistant that helps draft, edit, and refine proposals.

5. **Instrumentl** – AI-powered grant matching and tracking tool to find funding opportunities.

Marketing & Communications

6. **CAUSEWRITER.AI,** a FREE purpose-built AI tool for non-profits.
7. **Chatfuel for Nonprofits** – AI chatbot builder for Facebook Messenger to automate donor interactions.
8. **Copy.ai for Nonprofits** – Generates fundraising emails, social media posts, and donor thank-you letters.
9. **Lever for Good** – AI-driven storytelling tool that helps craft compelling narratives for campaigns.

Operations & Efficiency

9. **TABLEAU.COM -** Business Intelligence and Analytics Software
10. **Keela** – AI-powered CRM with smart donor insights, email automation, and fundraising analytics.
11. **Every.org AI** – Free fundraising platform with AI-driven donor recommendations.
12. **Bonterra (formerly CyberGrants)** – AI-powered grant management and impact measurement.

Impact Measurement & Reporting

12. **UpMetrics** – AI-driven impact analytics to track and visualize program outcomes.
13. **Soapbox Engage** – AI-powered advocacy and donation tracking for nonprofits.

www.ingramcontent.com/pod-product-compliance
Lightning Source LLC
Chambersburg PA
CBHW051755200326

41597CB00025B/4568